# Springer Tracts in Modern Physics 99

# Springer Tracts in Modern Physics

74 **Solid-State Physics** With contributions by G. Bauer, G. Borstel, H. J. Falge, A. Otto

75 **Light Scattering by Phonon-Polaritons** By R. Claus, L. Merten, J. Brandmüller

76 **Irreversible Properties of Type II Superconductors** By H. Ullmaier

77 **Surface Physics** With contributions by K. Müller, P. Wißmann

78 **Solid-State Physics** With contributions by R. Dornhaus, G. Nimtz, W. Richter

79 **Elementary Particle Physics** With contributions by E. Paul, H. Rollnick, P. Stichel

80* **Neutron Physics** With contributions by L. Koester, A. Steyerl

81 **Point Defects in Metals I:** Introductions to the Theory (2nd Printing)
By G. Leibfried, N. Breuer

82 **Electronic Structure of Noble Metals, and Polariton-Mediated Light Scattering**
With contributions by B. Bendow, B. Lengeler

83 **Electroproduction at Low Energy and Hadron Form Factors**
By E. Amaldi, S. P. Fubini, G. Furlan

84 **Collective Ion Acceleration** With contributions by C. L. Olson, U. Schumacher

85 **Solid Surface Physics** With contributions by J. Hölzl, F. K. Schulte, H. Wagner

86 **Electron-Positron Interactions** By B. H. Wiik, G. Wolf

87 **Point Defects in Metals II:** Dynamical Properties and Diffusion Controlled Reactions
With contributions by P. H. Dederichs, K. Schroeder, R. Zeller

88 **Excitation of Plasmons and Interband Transitions by Electrons** By H. Raether

89 Giant Resonance Phenomena in **Intermediate-Energy Nuclear Reactions**
By F. Cannata, H. Überall

90* **Jets of Hadrons** By W. Hofmann

91 **Structural Studies of Surfaces**
With contributions by K. Heinz, K. Müller, T. Engel, and K. H. Rieder

92 **Single-Particle Rotations in Molecular Crystals** By W. Press

93 **Coherent Inelastic Neutron Scattering in Lattice Dynamics** By B. Dorner

94 **Exciton Dynamics in Molecular Crystals and Aggregates** With contributions by
V. M. Kenkre and P. Reineker

95 **Projection Operator Techniques** in Nonequilibrium Statistical Mechanics
By H. Grabert

96 **Hyperfine Structure in 4d- and 5d-Shell Atoms** By S. Büttgenbach

97 **Elements of Flow and Diffusion Processes in Separation Nozzles** By W. Ehrfeld

98 **Narrow-Gap Semiconductors** With contributions by R. Dornhaus, G. Nimtz, and
B. Schlicht

99 **Dynamical Properties of IV–VI Compounds** With contributions by H. Bilz, A. Bussmann-
Holder, W. Jantsch, and P. Vogl

100* **Quarks and Nuclear Forces** Edited by D. C. Fries and B. Zeitnitz

101 **Neutron Scattering and Muon Spin Rotation** With contributions by R. E. Lechner,
D. Richter, and C. Riekel

---

* denotes a volume which contains a Classified Index starting from Volume 36.

# Dynamical Properties of IV–VI Compounds

Contributions by
H. Bilz   A. Bussmann-Holder   W. Jantsch   P. Vogl

With 47 Figures

Springer-Verlag Berlin Heidelberg GmbH 1983

Dr. Wolfgang Jantsch

Max-Planck-Institut für Festkörperforschung, Heisenbergstrasse 1
D-7000 Stuttgart 80, Fed. Rep. of Germany
*Permanent Address:* Johannes Kepler Universität Linz, Institut für Experimentalphysik,
A-4045 Linz-Auhof, Austria

Dr. Anette Bussmann-Holder
Professor Dr. Heinz Bilz

Max-Planck-Institut für Festkörperforschung, Heisenbergstrasse 1
D-7000 Stuttgart 80, Fed. Rep. of Germany

Dr. Peter Vogl

Universität Graz, Institut für Theoretische Physik,
A-8010 Graz, Austria

---

*Manuscripts for publication should be addressed to:*

Gerhard Höhler

Institut für Theoretische Kernphysik der Universität Karlsruhe
Postfach 6380, D-7500 Karlsruhe 1, Fed. Rep. of Germany

*Proofs and all correspondence concerning papers in the process of publication
should be addressed to:*

Ernst A. Niekisch

Haubourdinstrasse 6, D-5170 Jülich 1, Fed. Rep. of Germany

---

ISBN 978-3-662-13533-4      ISBN 978-3-540-39534-8 (eBook)
DOI 10.1007/978-3-540-39534-8

Library of Congress Cataloging in Publication Data. Main entry under title: Property of IV – VI compounds. (Springer tracts in modern physics; v. 99) Bibliography: p. 1. Semiconductors. I. Jantsch, W. (Wolfgang), 1946 –      . II. Title: Property of 4 – 6 compounds. III. Title: Property of four – six compounds. IV. Series. QCI.S797 vol. 99 [QC611]   539s   [537.6'22]   83-10299

# Preface

The IV-VI compounds such as PbS and SnTe form an unusual class of materials. They often show a soft-mode behaviour and exhibit characteristic dielectric anomalies. These properties reflect the fact that many IV-VI compounds are real or incipient ferroelectrics. Simultaneously, they are narrow-gap semiconductors with quite unique electronic properties.

This volume contains a review of the present experimental situation and of the theoretical understanding of the dynamical aspects of IV-VI compounds with emphasis on the ferroelectric properties.

In particular, the dielectric and optical measurements of several IV-VI compounds as a function of temperature, alloy concentration and carrier concentration provide an important insight into the driving mechanisms for the structural instability of these materials. This aspect is reviewed in the contribution by W. Jantsch.

The second contribution by A. Bussmann-Holder, P. Vogl and H. Bilz deals with the present theoretical description and interpretation of electronic and lattice-dynamical properties of IV-VI compounds. Three different microscopic models are discussed, which are used to explain the complex temperature dependence of soft modes and related properties. At present, it seems that the recently developed polarizability model is most successful in explaining a large variety of experimental data.

The present volume supplements the parallel review, by G. Nimtz and B. Schlicht, on IV-VI compounds which focuses on crystal preparation as well as transport and optical properties (Vol. 98 in this series).

It is hoped that our investigations turn out to be useful for both researchers and graduate students. They may be intrigued by the model character of these compounds which exhibit both structural simplicity and complex physical properties.

Linz,
Stuttgart, Graz, April 1983

*W. Jantsch*
*A. Bussmann-Holder*
*H. Bilz*
*P. Vogl*

# Contents

## Dielectric Properties and Soft Modes in Semiconducting (Pb, Sn, Ge) Te

By W. Jantsch (With 25 Figures)

1.  Introduction  .................................................................  1

2.  Experimental Determination of the Soft-Mode Frequency ................  5
    2.1  Neutron Scattering ...............................................  5
    2.2  Inelastic Tunneling of Electrons  ................................  7
    2.3  Raman Scattering .................................................  7
    2.4  Far-Infrared Spectroscopy  .......................................  11

3.  Experimental Determination of the Static Dielectric Constant  ...........  20
    3.1  Differential Capacitance Measurements  ...........................  20
    3.1  Microwave Techniques  ............................................  25

4.  Effects Related to the Phase Transition  ...............................  29
    4.1  Changes of Band Structure and Related Phenomena  .................  29
    4.2  Resistance Anomaly  ..............................................  30
    4.3  Acoustic and Specific-Heat Anomalies  ............................  31

5.  Results and Discussion  ...............................................  32
    5.1  Temperature Dependence of the Soft Mode and Phase Transition  .....  32
    5.2  Microscopic Origin of Structural Instability and Chemical Trends
         of Critical Temperature  .........................................  36
    5.3  Influence of Lattice Defects  ....................................  39
    5.4  Influence of Magnetic Fields  ....................................  44

6.  Summary ...............................................................  45

References  ...............................................................  46

Combined Subject Index ....................................................  99

# Electronic and Dynamical Properties of IV–VI Compounds

By A. Bussmann-Holder, H. Bilz, and P. Vogl (With 22 Figures)

1. Introduction ........................................................ 51
   1.1 History ......................................................... 51
   1.2 Landau Theory ................................................... 53

2. Chemical Structure and Electronic Bands of IV-VI Compounds .......... 54
   2.1 Structure and Ferroelectricity .................................. 54
   2.2 Ionicity and Covalency .......................................... 56
   2.3 Electronic Band Structure ....................................... 57
   2.4 The Zero-Gap Situation .......................................... 59

3. Lattice Dynamics and Phase Transitions .............................. 60
   3.1 General Aspects: The Soft-Mode Concept .......................... 60
   3.2 Electronic Theory of the Soft-Mode Instability .................. 61
   3.3 The Anharmonic Lattice Model .................................... 65
       3.3.1 Quasi-Harmonic Approximation ............................. 65
       3.3.2 The Molecular-Field Treatment of a Single Mode Model ...... 68
   3.4 The Vibronic Model .............................................. 71
   3.5 The Polarizability Model ........................................ 74
       3.5.1 Results .................................................. 81
       3.5.2 Three-Dimensional Models ................................. 85
   3.6 Comparison of Models ............................................ 87
   3.7 Nonlinear Excitations in IV-VI Semiconductors ................... 88

4. Summary and Conclusion .............................................. 93

References ............................................................. 94

Combined Subject Index ................................................. 99

# Dielectric Properties and Soft Modes in Semiconducting (Pb, Sn, Ge) Te

By W. Jantsch

## 1. Introduction

Binary compounds of Groups IV and VI of the periodic table have been of considerable theoretical, experimental and technological interest in recent years (Ravich
et al., 1970; Rabii, 1974; Rauluskiewicz et al., 1978). For the lead chalcogenides
PbTe, PbSe, PbS) in particular, the band structure (Dalven, 1973), ohmic (Ravich
et al., 1971) and high-field transport (Jantsch et al., 1978a), optical (Zemel et
al., 1965) and magnetooptical (Gerlach and Grosse, 1977; Bauer, 1978,1980) properties, and electronic states of defects (Heinrich, 1980; Lischka, 1982) have been
studied extensively.

The IV-VI compounds are narrow-gap semiconductors with a generally direct
minimum energy gap of typically 0.2 eV. Some alloy systems, like $Pb_{1-x}Sn_xTe$, exhibit a zero-gap transition depending on temperature and composition (Dimmock et
al., 1966). This property makes IV-VI compounds especially suitable for technological applications as infrared sources (Preier, 1979) and detectors (Holloway,
1980; Haas et al., 1975). Considerable effort has been devoted to the development
of crystal-growth techniques. Among these, the hot-wall epitaxy has proved to be
very successful for high-quality epitaxial material (Lopez Otero, 1978). In addition, this method enables a great deal of variation in growth conditions.

The IV-VI compounds exhibit a number of outstanding properties. Especially the
dielectric properties and structural transitions of the tellurides to a low-temperature ferroelectric phase have attracted considerable attention (Lines and Glass,
1977; Kawamura, 1978,1980). Less work has been done on the IV-VI compounds with
lower molecular weight (GeS, GeSe, SnS, SnSe) which crystallize in an orthorhombic
layer structure. Although phase transition may also occur for these compounds, the
present review concentrates on the tellurides because of their outstandingly simple
crystal structure and the comprehensive work published in the past few years.

First indications for an anomalously high dielectric constant of PbTe were derived from the observation of unusually high values of the carrier mobilities at
low temperatures (Allgaier and Scanlon, 1958; Allgaier and Houston, 1962). A differ

1

ent interpretation for the small contribution of ionized impurity scattering was given later (Pratt, 1973,1974). However, from measurements of the voltage-dependent capacity C(V) of Schottky barriers on PbTe, Kanai and Shohno (1963) obtained a static dielectric constant of 400 for temperatures below 130 K. In view of the connection between the dielectric properties and the lattice dynamics of ferroelectric crystals (Cochran, 1960), the high value of the static dielectric constant led Cochran (1964) to suggest a soft-mode behavior and the possibility of a displacive ferroelectric phase transition in PbTe.

The connection between the static dielectric constant $\varepsilon_0$ and the optical mode frequencies is evident from the LYDDANE-SACHS-TELLER relation (1941), given here in its extended version (Cochran and Cowley, 1962):

$$\frac{\varepsilon_0}{\varepsilon_\infty} = \prod_j \frac{\omega_{Lj}^2}{\omega_{Tj}^2} \quad . \tag{1.1}$$

Here $\varepsilon_\infty$ stands for the high-frequency dielectric constant and $\omega_{Lj}$, $\omega_{Tj}$, for the longitudinal and transverse optical frequencies of the $j^{th}$ phonon branch at $q = 0$. If one of the transverse modes becomes soft at a critical temperature $T_c$,

$$\omega_{Tj}(q = 0, T_c) = 0 \quad , \tag{1.2}$$

the crystal becomes unstable against this type of vibration. As a result, a lattice distortion occurs. For a diatomic cubic lattice of NaCl structure (the structure of PbTe), there is only one type of optical vibration, split into a doubly degenerate transverse and a singlet longitudinal mode. For $q = 0$, the two sublattices vibrate against each other. On cooling below $T_c$, the two sublattices are displaced with respect to each other along the direction of vibration of the soft mode: the soft mode "freezes". Due to the lowering of symmetry a permanent dipole moment occurs with eight possible orientations along the cubic <111> axes. The corresponding spontaneous polarization can be reoriented by an external electric field, which is the criterion of ferroelectricity. (This definition is strictly applicable only to insulators. In the present case, free carriers impede the application of sufficiently high electric fields.)

In SnTe (Pawley et al., 1966) and in PbTe (Alperin et al., 1972), $\omega_{TO}(q = 0)$ decreases with decreasing temperature. Measurements of the static dielectric constant using the C(V) method on graded p-n junctions (Bate et al., 1970) show that the static dielectric constant of PbTe obeys a Curie-Weiss law

$$\varepsilon_0^{-1} = C_0(T - T_c) \quad , \tag{1.3}$$

where $C_0$ is the temperature-independent Curie constant. For PbTe only negative values of $T_c$ have been found by extrapolation of $\varepsilon_0^{-1} \to 0$, indicating that PbTe remains paraelectric for all temperatures.

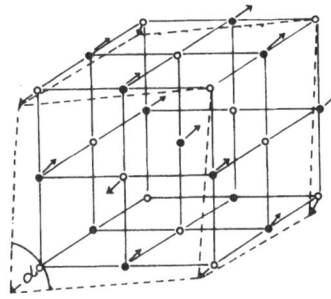

Fig.1.1. NaCl structure and rhombohedral distor-, tion of the low-temperature phase of binary Group-IV tellurides. The arrows indicate the sublattice shifts

Structural changes have been observed in GeTe (Schubert and Fricke, 1951,1953) and SnTe (Muldawer, 1973; Iizumi et al., 1975; Valassiades and Economou, 1975). At room temperature GeTe crystallizes in a rhombohedral ($C_{3v}$) structure, which may be viewed as a distorted NaCl structure: the angle $\alpha$ between the axes of the unit cell differs only slightly from 90° (Fig.1.1). With increasing temperature, the interaxis rhombohedral angle $\alpha$ of GeTe increases gradually from its room-temperature value of 88.35° to the cubic value of 90° (NaCl structure) at about 720 K (Schubert and Fricke, 1951,1953). For SnTe the same kind of structural transition has been observed (Muldawer, 1973; Iizumi et al., 1975). Samples with low carrier concentration ($p < 1 \cdot 10^{20}$ cm$^{-3}$) transform at about 100 K. With increasing carrier concentration the critical temperature decreases: the cubic structure is stabilized.

Alloy systems like $Sn_{1-x}Ge_xTe$ (Bierly et al., 1963), $Pb_{1-x}Sn_xTe$ (Shimada et al., 1977) and $Pb_{1-x}Ge_xTe$ (Kawamura, 1978; Jantsch et al., 1979) exhibit an increasing critical temperature with increasing x. X-ray analysis (Bierly et al., 1963; Schubert and Fricke, 1953) gives no indication of a discontinuity of the rhombohedral angle, which is an order parameter for this type of phase transition, at the critical temperature of SnTe or GeTe. The phase transition of $Sn_{1-x}Ge_xTe$, however, has recently been shown to become first order for x > 0.27 (Clarke, 1978). Results for the deviation of the rhombohedral angle from 90° ($\Delta\alpha = 90° - \alpha$) are given in Fig.1.2 as a function of temperature. For x > 0.27, $\Delta\alpha$ goes to zero discontinuously at the critical temperature.

Investigations of the Raman effect (Steigmeier and Harbecke, 1970; Sugai et al., 1977a,b), inelastic neutron scattering (Pawley et al., 1966; Alperin et al., 1972) and measurements of far-infrared reflectivity (Jantsch et al., 1978b) have shown that the structure change coincides with a softening of the transverse optical phonon mode, in agreement with the soft-mode concept (Cochran, 1960; Anderson, 1960).

In summary, the binary tellurides of Pb, Sn and Ge and their alloys form a class of semiconductors with a tendency to structural phase transitions caused by a softening of the zone-center TO mode. The phase transition in GeTe and SnTe is of the simplest conceivable type: it consists in a relative shift of the cation and anion sublattices. Compared to other "classical" ferroelectrics, the crystal structure

3

is particularly simple: the unit cell contains only two atoms. In alloy systems, the critical temperature varies within an exceedingly wide range: any value between 700 K and negative values can be attained by choice of the alloy composition. Some of these compounds seem to provide rare examples of the existence of displacive ferroelectric phase transitions of *second* order. These materials therefore offer a rare opportunity to test the soft-mode concept in the critical regime close to the phase transition. Such critical phenomena are of fundamental importance in under-standing phase transitions (Lines and Glass, 1977; Bruce and Cowley, 1980).

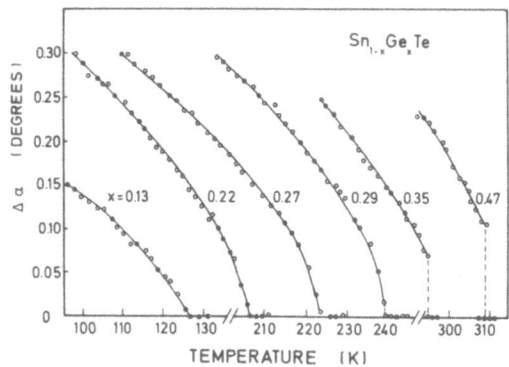

Fig.1.2. Deviation of the rhombo-hedral angle $\Delta\alpha$ from the cubic (90°) value of $Sn_{1-x}Ge_xTe$ as a function of temperature, after Clarke (1978)

The physical properties mentioned above, together with the technological impor-tance of the IV-VI compounds, make these materials interesting both in research and material science. Presumably they will also become standard materials in the field of ferroelectricity.

Complications arise, however, because IV-VI compounds are narrow-gap semiconduc-tors with high concentrations of highly mobile carriers. Therefore most of the standard techniques used to investigate soft modes and dielectric properties have to be modified for application to this class of materials.

In Chaps.2 and 3, experimental methods for the investigation of soft modes and the static dielectric constant, respectively, are reviewed. Chapter 4 discusses effects related to the phase transition, some of which, especially an anomaly of the elec-trical resistivity, have been used as simple tools to determine the critical temperature.

Finally, in Chap.5 experimental results for the dielectric behavior and soft modes are discussed in the light of existing theoretical models. The emphasis is not on completeness or a detailed description of theoretical aspects, but rather on a sur-vey of this field.

## 2. Experimental Determination of the Soft-Mode Frequency

According to the Cochran-Anderson (1960) theory, ferroelectric phase transitions are caused by "softening" of the zone-center TO-phonon mode. Four methods have been applied to investigate experimentally $\omega_{TO}(q = 0)$:

 i)   inelastic neutron scattering
 ii)  inelastic tunneling of electrons
 iii) Raman spectroscopy
 iv)  far-infrared spectroscopy.

These methods are discussed below in the context of their application to IV-VI compounds.

### 2.1 Neutron Scattering

In contrast to other methods, inelastic neutron scattering experiments enable a determination of the complete, wave-vector-dependent dispersion curves, thus allowing a comparison with lattice dynamical models (Cochran et al., 1966). However, this method requires large crystals of good quality and homogeneity, which, especially in the case of alloys of IV-VI compounds, are difficult to grow. Therefore, only a limited amount of data is available in the literature. The first evidence for mode softening on cooling in the paraelectric phase was established by Pawley et al. (1966) for SnTe (Figs.2.1,2). With decreasing temperature, $\omega_{TO}(q \rightarrow 0)$ decreases. At low temperatures, $\omega_{TO}(q \rightarrow 0)$ saturates and no phase transition occurs for this particular sample. Complete dispersion curves for PbTe at 296 K and for SnTe at 100 K were measured by Cochran et al. (1966) and Cowley et al. (1969), respectively. For SnTe, which exhibits carrier concentrations as high as $10^{20}$ cm$^{-3}$ due to intrinsic defects, the LO branch is expected to approach the TO branch for $q \rightarrow 0$ since the macroscopic electric field of the LO mode is screened by free carriers (R.A. Cowley and Dolling, 1965; E.R. Cowley et al., 1969), (Fig.2.1). Table 2.1 summarizes references of available data on inelastic neutron scattering.

No data has been published for the low-temperature phase of any member of the Group IV telluride family as yet. However, from measurements of the temperature dependence of elastic neutron scattering on $Pb_{1-x}Sn_xTe$ and SnTe with low defect concentration, evidence for a displacive phase transition was reported by Komatsubara et al. (1974) and Iizumi et al. (1975), respectively. In this method, the Bragg peak intensities of odd-index reflections, which are very weak in the cubic NaCl structure, are measured as a function of temperature. The phase transition manifests itself by an increase of the "forbidden" Bragg intensities due to the lowering of symmetry. The sublattice shift, which is an order parameter, can be evaluated from the normalized intensity. The results indicate that the ferroelectric phase transitions of SnTe and $Pb_{1-x}Sn_xTe$ are of second order.

**Table 2.1.** References for inelastic neutron-scattering investigations

| Material | Temp. [K] | Measured phonon branch | Reference |
|---|---|---|---|
| SnTe | 6–300 | TO,LO(q‖[001]) | Pawley et al. (1966) |
| SnTe | 100 | complete | E.R. Cowley et al. (1969) |
| PbTe | 4,293<br>4–293 | TO,LO(q‖[111])<br>TO(q → 0) | Alperin et al. (1972) |
| PbTe | 296 | LO([001], [110], [111]) | R.A. Cowley and Dolling (1965) |
| PbTe | 296 | complete | Cochran et al. (1966) |
| $Pb_{0.2}Sn_{0.8}Te$ | 94 | TO, A | |
| $Pb_{0.63}Sn_{0.37}Te$ | 4.2 | TO, A | Dolling and Buyers (1973) |
| $Pb_{0.63}Sn_{0.37}Te$ | 4–220 | TO(q → 0) | |
| $Pb_{0.87}Sn_{0.13}Te$ | 19–293 | TO(q → 0), linewidth | Daughton et al. (1978) |
| $Pb_{0.8}Sn_{0.2}Te$ | 5–239 | TO(q → 0), linewidth | |
| $Pb_{0.8}Sn_{0.2}Se$ | 80 | TO,LO,LA,TA(q‖[001]) | |
| $Pb_{0.8}Sn_{0.2}Se$ | 5–250 | TO(q → 0) | Vodopyanov et al. (1978) |
| $Pb_{0.93}Sn_{0.07}Se$ | 5–250 | TO(q → 0), linewidth | |

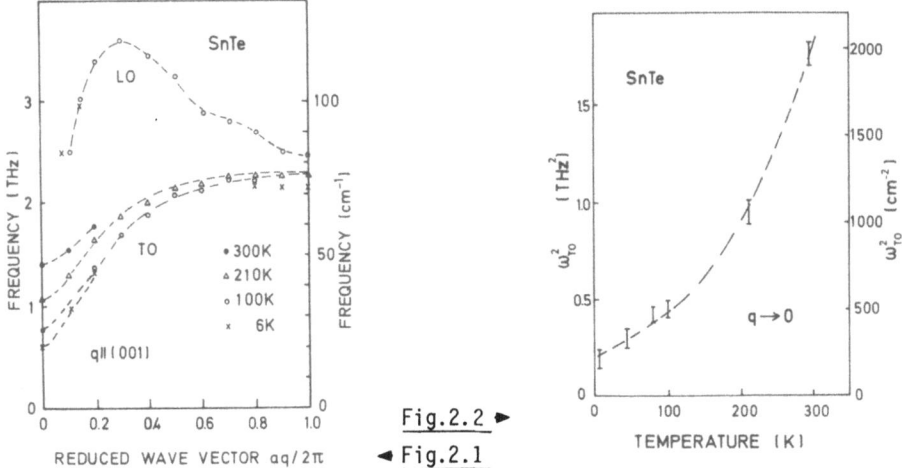

Fig.2.2 ►

◄ Fig.2.1

**Fig.2.1.** Dispersion of the LO and TO modes of SnTe at various temperatures obtained from inelastic neutron scattering, after Pawley et al. (1966)

**Fig.2.2.** Squared soft-mode frequency [$\omega_{TO}(q \to 0)$] from Fig.2.1, after Pawley et al. (1966)

## 2.2  Inelastic Tunneling of Electrons

Phonon-assisted tunneling was investigated for p-type $Pb_{1-x}Sn_xTe$ at liquid helium temperatures by Takasaki and Tanaka (1977). Tunneling junctions were prepared as metal-insulator-semiconductor (MIS) structures of the type $Pb-SiO_2$ (or $ZnS$)-$Pb_{1-x}Sn_xTe$. Peaks observed in the second derivative of the current-voltage characteristics, $\partial^2 I/\partial V^2$, versus bias voltage were attributed to the TO and LO modes at the $\Gamma$ and L points of the Brillouin zone. The energy gap can be determined also from this kind of experiment. With increasing x, a decrease of the energy gap and a decrease of $\omega_{TO}(q \rightarrow 0)$ occurs. With increasing carrier concentration, $\omega_{TO}(q \rightarrow 0)$ increases. These results were interpreted in terms of the interband electron-phonon coupling model by Kawamura et al. (1975). For PbTe the carrier concentration was not found to influence the LO-phonon energy.

## 2.3  Raman Scattering

First-order Raman scattering in Group IV tellurides is restricted to the low-temperature rhombohedral phase, since for the NaCl structure $(T > T_c)$ all the atoms are situated at centers of inversion. In the latter situation the optical phonon modes are IR active but Raman inactive.

The optical phonons in the cubic NaCl phase ($O_h$ structure) belong to a 3-fold degenerate $F_{1u}$ mode, whose degeneracy is lifted by the macroscopic electric field associated with the longitudinal mode. In the rhombohedral phase ($C_{3v}$ structure), the $F_{1u}$ mode is split into a doubly degenerate E- and a single $A_1$ mode. The eigenvector of the $A_1$ mode corresponds to the static deformation due to the rhombohedral distortion. It is this mode which is responsible for the phase transition (Steigmeier and Harbeke, 1970).

Both the E- and the $A_1$ modes are IR- and Raman active in the $C_{3v}$ structure. The Raman tensors are given by (Loudon, 1964):

$$T(A_{1z}) = \begin{pmatrix} a & 0 & 0 \\ 0 & a & 0 \\ 0 & 0 & b \end{pmatrix}, \; T(E_y) = \begin{pmatrix} c & 0 & 0 \\ 0 & -c & d \\ 0 & d & 0 \end{pmatrix}, \; T(E_{-x}) = \begin{pmatrix} 0 & -c & d \\ -c & 0 & 0 \\ -d & 0 & 0 \end{pmatrix}, \quad (2.1)$$

where the extraordinary $A_1$ mode is polarized in the rhombohedral c-direction ($\equiv$ z-axis) and the ordinary E modes in the x-y plane. For "diagonal" polarization [same polarization of incident and scattered beam, e.g., $\bar{z}(xx)z$ using the notation of Damen et al., 1966] both the $A_1$- and E modes can be observed, whereas for crossed polarization [e.g., $\bar{z}(yx)z$] only the E modes give a nonvanishing contribution.

The interpretation of Raman spectra of ferroelectric IV-VI compounds may be complicated by several effects:

(i) Due to the high absorption coefficient of these materials in the visible range, wave vector conservation is not strictly fulfilled. Therefore the resulting line shape is expected to contain contributions from a finite volume in q space rather than the q → 0 component of the dispersion curves (Abstreiter et al., 1979; Katayama and Mills, 1979). The observed peak position in this case does not coincide with the phonon frequency and the line width is enhanced with respect to the pure phonon line width at $q = 0$.

(ii) In the presence of high concentrations of free carriers, coupled LO-phonon-plasmon modes instead of pure phonon lines and single-particle excitation spectra may occur. In both cases the Raman spectra depend on the carrier concentration and on the scattering wave vector. Both effects are sensitive to surface states, since surface states may cause a band bending and hence gradients in the carrier concentration.

(iii) In the rhombohedral phase the crystal divides into ferroelectric domains (Snykers et al., 1972; Jantsch et al., 1979,1981b; Lewis et al., 1980): the dipole moment of an individual domain is oriented along one of the four possible directions corresponding to the cubic <111> axes. The domain size is of the order of 1 μm (Snykers et al., 1972; Jantsch et al., 1981b). Therefore the usual selection rules cannot be used to identify the various modes.

First-order Raman scattering has been observed in GeTe (Steigmeier and Harbeke, 1970), SnTe (Brillson et al., 1974; Murase et al., 1976; Sugai et al., 1977a,b; Murase and Sugai, 1979) and in $Pb_{1-x}Ge_xTe$ (Sugai et al., 1979). The experiments were performed in backscattering geometry in the visible range. Due to the high absorption coefficient, Raman scattering is restricted to a surface region extending to a depth of about 100 Å. Consequently, the Raman intensities are very low (30-100 counts/s for typical experimental conditions) and the spectra are susceptible to the surface preparation method (Brillson and Burstein, 1971; Shimada et al., 1977; Cape et al., 1977). Frequently an iodine filter is used with the 5145 Å argon-ion laser line to suppress the strong elastic component, which otherwise obscures the low-frequency phonon lines (Devlin et al., 1971).

In SnTe and $Pb_{0.95}Ge_{0.05}Te$ two lines have been observed below 50 $cm^{-1}$ whose frequencies decrease on approaching the critical temperature from below (Figs.2.3,4). These lines do not occur above the critical temperature. In $Pb_{0.95}Ge_{0.05}Te$ a directional dispersion of the upper line was also found (Fig.2.3). Therefore the upper line has been assigned to the extraordinary $A_1$ mode. For k‖[111], (using pseudo-cubic indices) the $A_1$ mode is observable only due to the presence of ferroelectric domains, whose c-axis is not parallel to the laser beam. In later experiments on $Pb_{1-x}Ge_xTe$, the $A_1$-E mode splitting could not be resolved (Jantsch and Stolz, 1980). Typical Raman spectra obtained in backscattering geometry ($\bar{z}(xx)z$) are given in Fig.2.5. The linewidth is comparable to that of each of the two lines in Fig.2.3 but much larger than the linewidth obtained from far-infrared spectroscopy (Sect. 2.4), which can be attributed to effects of the extremely small penetration depth (see above). Mode splitting is discussed in Sect.2.4.

With increasing carrier concentration, a decrease in frequency of a single line at about the E-mode frequency has been reported by Sugai et al. (1977b) (Fig.2.6).

8

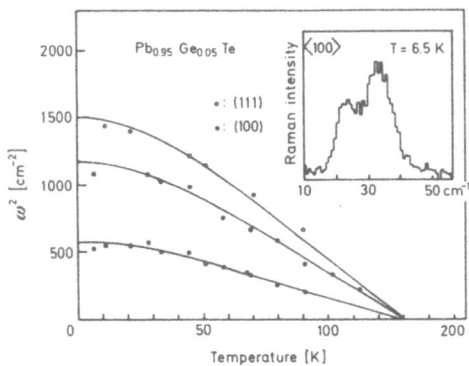

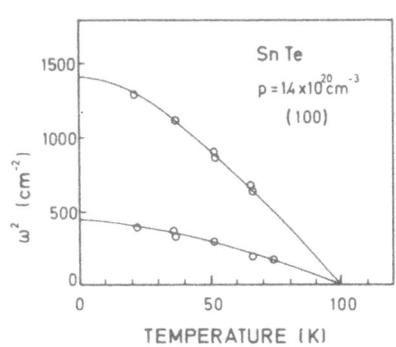

**Fig.2.3.** Temperature dependence of squared Raman line frequencies observed for (111) and (100) surfaces, respectively, of $Pb_{1-x}Ge_xTe$ (x = 0.05) in the rhombohedral phase. The inset shows a typical Raman spectrum, after Murase and Sugai (1979)

**Fig.2.4.** Temperature dependence of squared Raman line frequency observed on a SnTe (100) surface in the rhombohedral phase, after Murase and Sugai (1979)

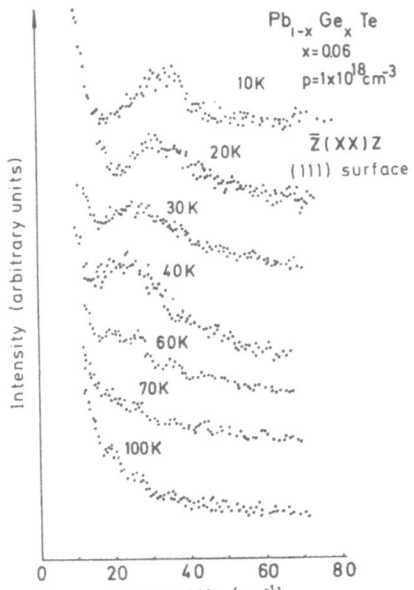

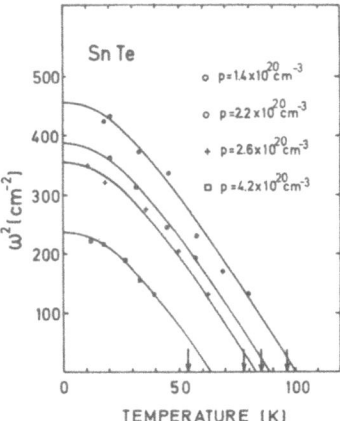

**Fig.2.6.** Temperature dependence of squared Raman line frequencies observed for p-SnTe with different hole concentrations. Scattering is probably due to the $E_{TO}$ mode, after Sugai et al. (1977b)

**Fig.2.5.** Raman spectra ($\bar{z}(xx)z$) of a $Pb_{0.94}Ge_{0.06}Te$ film grown epitaxially on a (111) oriented $BaF_2$ substrate. For crossed polarization ($\bar{z}(xy)z$) the spectra are essentially unchanged, after Jantsch and Stolz (1980)

Softening of a mode in the low-temperature phase corresponds to decreasing critical temperature, i.e., a stabilization of the cubic phase due to free carriers or intrinsic lattice point defects, which are the origin of free carriers in undoped material. For a coupled plasmon-phonon polariton, one would expect an increase in frequency

9

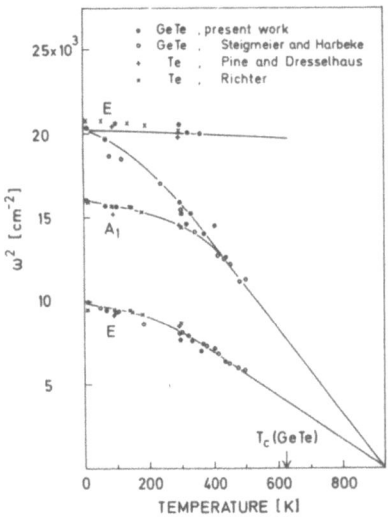

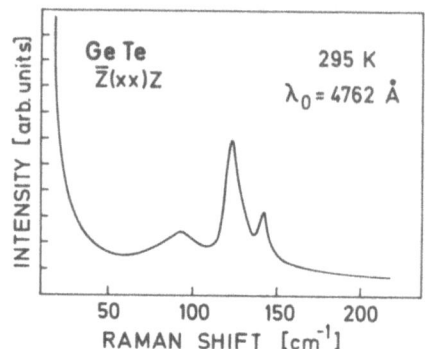

Fig.2.8. Typical Raman spectrum of GeTe at 295 K in backscattering geometry (Jantsch and Stolz, 1980)

Fig.2.7. Squared Raman line shifts as observed in GeTe [from Steigmeier and Harbeke (1970) and Jantsch and Stolz (1980)] and in Te [from Pine and Dresselhaus (1971) and Richter (1973)]. The arrow indicates the critical temperature of GeTe obtained from differential thermal analysis and X-ray investigations (Steigmeier and Harbeke, 1970)

with increasing carrier concentration, in contrast to the experimental result. Therefore this effect (Sugai et al., 1977b) is either overcompensated by the stabilization of the phonon mode or only the pure unscreened phonon modes are observed.

For GeTe two Raman lines have been observed (Steigmeier and Harbeke, 1970), about a factor of 3 higher in frequency than in the case of SnTe or $Pb_{0.95}Ge_{0.05}Te$ (Fig.2.7). In contrast to the results for SnTe and $Pb_{1-x}Ge_xTe$, selection rules depending on the relative polarization of laser and Raman light allowed an assignment of these lines to the $A_1$- and E modes. The observation of selection rules indicates that only a single domain was involved in this experiment. The conditions for preparing large or even single domain crystals have not been discussed. For $k \parallel c$ and $k \perp c$ the same results with respect to the line position and its temperature dependence were obtained (Harbeke and Steigmeier, 1980) thus indicating that the observed TO- and LO modes are degenerate. The splitting due to the macroscopic field of the LO mode is perfectly screened by the free carriers, whose concentration is of the order of $10^{20}$ cm$^{-3}$ in SnTe and GeTe.

In Fig. 2.7 recent results obtained on GeTe are given together with results for pure Te. The close similarity suggests that a Te surface-layer was present in the experiment on GeTe (Jantsch and Stolz, 1980): Te surface-layers have also been observed on CdTe (Zitter, 1971) and $Pb_{1-x}Sn_xTe$ (Cape et al., 1977) by Raman spectroscopy. The observed Raman spectra are very similar to those obtained for GeTe given in Fig.2.8. A linear extrapolation of $\omega_{TO}^2 \rightarrow 0$ yields a critical temperature far above the value obtained from X-ray investigations (Schubert and Fricke, 1951;

Table 2.2. References for available Raman data

| Material | Investigated Temp. range [K] | Carrier concentration [cm$^{-3}$] | References |
|---|---|---|---|
| GeTe | 55-480 | | Steigmeier and Harbeke (1970) |
| SnTe | 3-80 | $1.5 \cdot 10^{20}$ | Brillson et al. (1974) |
| SnTe | 20-80 | $1.1 \cdot 10^{20}$ | Murase et al. (1976) |
| SnTe | 20-80 | $1.1 \cdot 10^{20}$ | Sugai et al. (1977a) |
| SnTe | 15-80 | $1.4\text{-}4.2 \cdot 10^{20}$ cm$^{-3}$ | Sugai et al. (1977b) |
| SnTe | 15-80 | $1.4\text{-}4.2 \cdot 10^{20}$ cm$^{-3}$ | Kawamura et al. (1978) |
| SnTe | 10-300 | $1.9\text{-}7.8 \cdot 10^{20}$ | Shimada et al. (1977) |
| $Pb_{1-x}Sn_xTe$ ($0 \leq x \leq 0.54$) | 10-300 | | Shimada et al. (1977) |
| $Pb_{1-x}Ge_xTe$ ($x = 0.05$) | 5-130 | $4.2 \cdot 10^{18}$ | Murase et al. (1979) |
| $Pb_{1-x}Ge_xTe$ ($x = 0.05$) | 5-130 | $4.2 \cdot 10^{18}$ | Sugai et al. (1979) |

Steigmeier and Harbeke, 1970). This discrepancy may also indicate that the Raman spectra observed on GeTe surfaces are actually due to Te.

A summary of available references on Raman scattering in Group IV tellurides is given in Table 2.2. So far, a detailed analysis of line shapes and scattering cross sections has not been performed, probably because of the experimental difficulties associated with this kind of material.

## 2.4 Far-Infrared Spectroscopy

Far-infrared (FIR) spectroscopy as a method for investigating the soft-mode frequency of IV-VI compounds has some advantages compared to the Raman effect. First of all, the optical phonons are also infrared (IR) active in the cubic phase, whereas first-order Raman scattering is restricted to the low-temperature rhombohedral phase. Surface effects are less important, since the light penetration depth in FIR is much larger than in Raman experiments, the latter being carried out in the visible range far above the fundamental absorption edge. Sample inhomogeneities and space charge layers can be recognized and ruled out (Tennant and Cape, 1976).

Evaluation of phonon-mode parameters from the FIR optical properties is somewhat complicated, since a model for the dielectric function is required. For typical samples, the plasma frequency, $\omega_p$, is above the Reststrahlen regime. Thus the transmission is very low and only the reflectivity can be measured with sufficient accuracy. The reflectivity is governed by the complex dielectric function in the frequency range considered. In principle, the dielectric function could be evaluated from the reflectivity using the Kramers-Kronig relations. This method requires, however, a high photometric accuracy, which is difficult to achieve in the FIR. Therefore the reflectivity is calculated from a model dielectric function, which contains phenomenological phonon- and plasmon-parameters. These are determined by fitting the model reflectivity to the experimental data.

The accuracy of this method can be improved significantly by using epitaxial films of several $\mu$m thickness, comparable to the light penetration depth in the FIR. Due to the high dielectric constant and the high carrier concentrations, bulk samples exhibit nearly total reflectivity in this spectral regime. The characteristic structures, caused by phonons, etc., are rather small. In thin film samples, however, multiple internal reflection and interference effects occur, which tend to enhance small dips in the reflectivity spectra. In the model calculation, multiple interference within the film and the substrate must be taken into account. The dielectric function of the substrate can be determined from the same kind of experiment.

The dielectric function, $\tilde{\varepsilon}(\omega)$, of PbTe can be described by a conventional dispersion oscillator model including a Drude term for the free carriers (Perkowitz, 1975; Burkhard et al., 1976):

$$\tilde{\varepsilon}(\omega) = \varepsilon_\infty + \tilde{\chi}_{ph} + \tilde{\chi}_{FC} = \varepsilon_\infty + \frac{\Delta\varepsilon \cdot \omega_{TO}^2}{\omega_{TO}^2 - \omega^2 - i\omega\Gamma} - \frac{\omega_p^2}{\omega(\omega + i\omega_\tau)} \; . \tag{2.2}$$

Here $\omega$ denotes the photon frequency, $\omega_{TO}$ the TO-phonon frequency and $\Delta\varepsilon = \varepsilon_0 - \varepsilon_\infty$ the difference between the static dielectric constant and the high-frequency dielectric constant. The latter takes into account contributions from short wavelength excitations; $\Gamma$ and $\omega_\tau$ are lattice- and free carrier-damping parameters, respectively, and $\omega_p$ stands for the unscreened plasma frequency

$$\omega_p^2 = \frac{4\pi N e^2}{m_p^*} \tag{2.3}$$

(N: carrier concentration; $m_p^* = m_t^* \cdot 3K/(2K+1)$: plasma effective mass, where $m_t^*$ is the transverse effective mass and $K = m_\ell^*/m_t^*$ the effective-mass anisotropy). The phonon contribution is frequently written also in the following form:

$$\tilde{\chi}_{ph} = \varepsilon_\infty \cdot \frac{\omega_{LO}^2 - \omega_{TO}^2}{\omega_{TO}^2 - \omega^2 - i\omega\Gamma} \; , \tag{2.4}$$

or

$$\tilde{\varepsilon}_{ph} = \tilde{\chi}_{ph} + \varepsilon_\infty = \varepsilon_\infty \cdot \frac{\omega_{LO}^2 - \omega^2 - i\omega\Gamma}{\omega_{TO}^2 - \omega^2 - i\omega\Gamma} \; , \tag{2.5}$$

where the LYDDANE-SACHS-TELLER (1946) relation has been used

$$\varepsilon_0/\varepsilon_\infty = \omega_{LO}^2/\omega_{TO}^2 \; . \tag{2.6}$$

From the model dielectric function the reflectivity and transmission are calculated for the film-substrate sandwich:

$$R_{123} = \left\{ R_{12}e^{\beta} + R_{23}e^{-\beta} + 2(Re\{r_{12}\}\cdot Re\{r_{23}\} + Im\{r_{12}\}\cdot Im\{r_{23}\})\cos\delta + \right.$$

$$\left. + 2(Re\{r_{12}\}\cdot Im\{r_{23}\} - Re\{r_{23}\}\cdot Im\{r_{12}\})\sin\delta \right\}/D \tag{2.7}$$

and

$$T_{123} = T_{12}T_{23}T_{31}e^{-\eta}/D \quad , \tag{2.8}$$

where

$$D = e^{\beta} + R_{12}R_{23}e^{-\beta} + 2(Re\{r_{12}\}\cdot Re\{r_{23}\} - Im\{r_{12}\}\cdot Im\{r_{23}\})\cos\delta + $$

$$+ 2(Re\{r_{12}\}\cdot Im\{r_{23}\} + Re\{r_{23}\}\cdot Im\{r_{12}\})\sin\delta \tag{2.9}$$

and

$$R_{jk} = |r_{jk}|^2 \quad , \quad r_{jk} = \frac{\tilde{n}_j - \tilde{n}_k}{\tilde{n}_j + \tilde{n}_k} \quad , \quad \tilde{n}_j = \sqrt{\tilde{\epsilon}_j(\omega)} = n_j - i\kappa_j$$

with $j = 1$ (vacuum), 2 (film), 3 (substrate), and $T_{jk} = (r_{jk} - 1)\cdot(r_{jk}^* - 1)$; $\delta = 2n_2\omega d_2/c$; $\beta = 2\kappa_2\omega d_2/c$; and $\eta = 2\kappa_3\omega d_3/c$. Here the film and the substrate are characterized by their complex refractive indices $\tilde{n}_2$ and $\tilde{n}_3$ and their thicknesses $d_2$ and $d_3$, respectively. In (2.7) multiple interference within the substrate is not taken into account. The optical path length within the substrate (thickness $d_3 \simeq 1$ mm) is large compared to that in the film. Interference effects due to a perfectly parallel substrate thus introduce a modulation of much higher periodicity in the frequency domain. The complete expression for this case has been given by Burkhard et al. (1977). Using wedged substrates or lower wavelength resolution, this additional complication is avoided.

Figure 2.9a illustrates this method with the aid of results from the dispersion oscillator model for the real part of the dielectric function $Re\{\tilde{\epsilon}(\omega)\}$. The parameters used (given in Fig.2.9b) are typical for PbTe. The dashed curve gives the free-carrier contribution, the dotted one the contribution from the phonon oscillator. The solid curve represents the total dielectric function, that is, the sum of both susceptibility contributions plus the high-frequency dielectric constant. Arrows indicate the mode frequencies $\omega_{TO}$, $\omega_{LO}$, $\omega_p^*$ and $\omega_L^+$, where

$$\omega_p^* = \omega_p/\sqrt{\epsilon_\infty} \tag{2.10}$$

is the plasma frequency screened by the valence electrons and

$$\omega_L^+ \simeq \sqrt{\omega_{LO}^2 + \omega_p^{*2}} \tag{2.11}$$

is the upper branch frequency of the coupled plasmon - phonon mode at $q \simeq 0$.

13

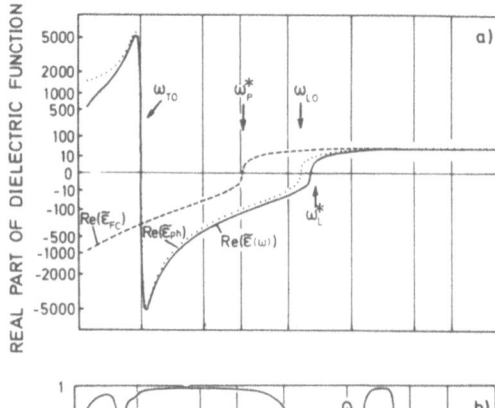

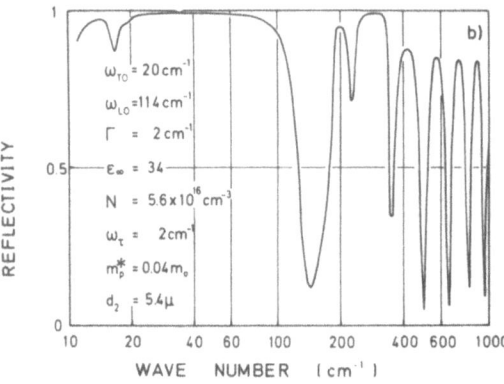

REAL PART OF DIELECTRIC FUNCTION

REFLECTIVITY

WAVE NUMBER (cm⁻¹)

a)

$\omega_{TO}$  $\omega_P^*$  $\omega_{LO}$

Re($\tilde{\varepsilon}_{FC}$)  Re($\tilde{\varepsilon}_{ph}$)  Re($\tilde{\varepsilon}(\omega)$)

$\omega_L^*$

b)

$\omega_{TO} = 20\,\text{cm}^{-1}$
$\omega_{LO} = 114\,\text{cm}^{-1}$
$\Gamma = 2\,\text{cm}^{-1}$
$\varepsilon_\infty = 34$
$N = 5.6 \times 10^{16}\,\text{cm}^{-3}$
$\omega_\tau = 2\,\text{cm}^{-1}$
$m_p^* = 0.04\,m_o$
$d_2 = 5.4\,\mu$

**Fig.2.9.** (a) Model calculation for the real part of the dielectric function of PbTe due to the optical phonons $\mathrm{Re}\{\varepsilon_{ph}\} = \mathrm{Re}\{\tilde{x}_{ph}\} + \varepsilon_\infty$ (dots), the free carriers $\mathrm{Re}\{\varepsilon_{FC}\} = \mathrm{Re}\{\tilde{x}_{FC}\} + \varepsilon_\infty$ (dashed line), and the total dielectric function $\mathrm{Re}\{\tilde{\varepsilon}\} = \mathrm{Re}\{\tilde{x}_{FC} + \tilde{x}_{ph}\} + \varepsilon_\infty$ (solid line). A nonlinear scale ($\varepsilon^{1/3}$) has been chosen to emphasize the zeros of the various contributions at $\omega_p^*$, $\omega_{LO}$ and $\omega_L^+$. (b) Reflectivity for a film(PbTe)-substrate (BaF$_2$) sandwich taking into account multiple interference in the PbTe film. The parameters used are given

In Fig.2.9b the reflectivity resulting from (2.5) is given as a function of frequency. For the substrate the parameters of BaF$_2$,

$$\Delta\varepsilon = 5.04, \quad \varepsilon_\infty = 2.16 \ , \quad \omega_{TO} = 185\ \text{cm}^{-1} \ , \quad \Gamma = 3\ \text{cm}^{-1} \ ,$$

have been used. As can be seen from Fig.2.9b, the reflectivity is very high below the plasma edge ($\omega < \omega_L^+$). Close to the TO-phonon resonance, the reflectivity drops due to the large peak in the phonon contribution to the real part of the dielectric function. The high-frequency onset of this structure approximately coincides with the TO-mode frequency, $\omega_{TO}(q \to 0)$, indicated by an arrow in Fig.2.9a. The additional structure at lower frequencies is due to interference effects within the film.

At the plasma edge, $\omega \approx \omega_L^+$, the real part of $\tilde{\varepsilon}(\omega)$ becomes positive and the thin film sample becomes nearly transparent. The reflectivity exhibits oscillations due to Fabry-Perot-type interference effects. Additional structure in this range originates from the Reststrahlen band of the BaF$_2$ substrate.

To fit the oscillator model to the observed reflectivity spectra in the cubic phase, the following parameters are required:

$\varepsilon_\infty$       : high-frequency dielectric constant

$\Delta\varepsilon = \varepsilon_0 - \varepsilon_\infty$ : phonon oscillator strength

$\omega_{TO}$      : soft-mode frequency

$\Gamma$ : phonon damping parameter

$\omega_p$ : plasma frequency

$\omega_\tau$ : free-carrier damping.

The high-frequency dielectric constant is obtained from the extrapolation of the square of the refractive index towards low frequencies ($\omega_L^+ < \omega << E_g/\hbar$). The refractive index can be evaluated from the periodicity of the interference fringes below the fundamental absorption edge ($\hbar\omega < E_g$) (Jantsch, 1980). Literature values for $\varepsilon_\infty$ are given in Table 2.3. An anomaly of the temperature dependence of $\varepsilon_\infty$ at the phase transition is discussed in Sect.4.1.

Table 2.3. Optical dielectric constant $\varepsilon_\infty$ of (PbSnGe)Te

| | x | Carrier conc. [cm$^{-3}$] (<0 for n type) | T[K] | $\varepsilon_\infty$ | Remarks | References |
|---|---|---|---|---|---|---|
| Pb$_{1-x}$Sn$_x$Te | 0 | $5 \cdot 10^{17}$ | 373 | 32.6 | | Zemel et al. |
| | | | 300 | 33.4 | | (1965) |
| | | | 77 | 38.5 | | |
| | 0.06 | $4.66 \cdot 10^{17}$ | 300 | 34.5 | Dependence on | Dionne and |
| | | | 82 | 38.8 | temperature | Woolley |
| | | | | | (84-300 K) and | (1972) |
| | 0.208 | $2.56 \cdot 10^{18}$ | 300 | 40.5 | carrier concen- | |
| | | | 82 | 43.3 | tration | |
| | | | | | ($5 \cdot 10^{17} - 10^{20}$ | |
| | | | | | cm$^{-3}$) investi- | |
| | | | | | gated | |
| | 0 | $-8.2 \cdot 10^{16}$ | 300 | 33 | | Lowney and |
| | | | 80 | 37 | | Senturia |
| | | | 6 | 39 | | (1976) |
| | 0.24 | $9.8 \cdot 10^{17}$ | 300 | 45 | | |
| | | | 80 | 53 | | |
| | | | 6 | 55 | | |
| | 0.36 | $4.8 \cdot 10^{18}$ | 300 | 50 | | |
| | | | 6 | 59 | | |
| | 0.39 | $8.0 \cdot 10^{17}$ | 300 | 50 | | |
| | | | 80 | 60 | | |
| | | | 6 | 60 | | |
| | 0.4 | $7 \cdot 10^{18}$ | 77 | 59 | | |
| | | | 6 | 59 | | |
| | 1 | $5 \cdot 10^{19}$ | 300 | 50 | Dependence on carrier con- centration ($5 \cdot 10^{19} - 5 \cdot 10^{20}$ cm$^{-3}$) | Riedl et al. (1967) |
| | 1 | $3.5 \cdot 10^{20}$ | 300 | 42 | Dependence on carrier con- centration ($3.5 \cdot 10^{20} - 1.4 \cdot 10^{21}$) | Finkenrath and Köhler (1966) |

Table 2.3 (cont.)

| | x | Carrier conc. [cm$^{-3}$] (<0 for n type) | T[K] | $\varepsilon_\infty$ | Remarks | References |
|---|---|---|---|---|---|---|
| $Pb_{1-x}Sn_xTe$ | 1 | $2 \cdot 10^{19}$ | 300 | 52 | Dependence on carrier concentration $(2.1 \cdot 10^{19} - 4.5 \cdot 10^{20})$ | Ota and Rabii (1971) |
| | 1 | $1.3 \cdot 10^{20}$ | 300 | 53 | Dependence on temperature (4.2–300 K) | Murase and Sugai (1979) |
| | 0 | $5.6 \cdot 10^{16}$ | 300 | 34.2 | Dependence on temperature (4.2–300 K) | Murase and Sugai (1979) |
| $Pb_{1-x}Ge_xTe$ | 0.07 | $4.7 \cdot 10^{18}$ | 300 | 39 | Dependence on temperature (4.2–300 K) | Murase and Sugai (1979) |
| | 1 | - | 300 | 36 | E polarization | Burstein et al. (1971) |

The amplitude of the interference fringes is very sensitive to the carrier damping parameter $\omega_\tau$. For an optimum fit, $\omega_\tau$ turns out to be frequency dependent. A detailed analysis of the frequency dependence of the free-carrier damping parameter of PbTe has been carried out by Burkhard et al. (1978). At low temperatures a strong resonance-like increase of the electron damping parameter was observed in the region of the coupled LO-phonon-plasmon excitations. This behavior can be explained quantitatively by the Mycielski (1978) model, which considers absorption of electromagnetic radiation from interaction of free carriers and collective plasma oscillations in a nonperfect, polar crystal. At higher temperatures the peak in $\omega_\tau$ disappears due to different screening and thermal broadening and only a smooth variation of $\omega_\tau$ with frequency is required to fit the reflectivity spectra (Burkhard et al., 1976,1978).

The plasma frequency $\omega_p$ can be calculated from (2.3) using d.c. Hall data for the carrier concentration and effective mass values as determined from magneto-optical investigations (Bauer, 1978). Using (2.6), the phonon oscillator strength in (2.2) can be replaced by the phonon frequencies, giving (2.4 or 2.5).

The longitudinal mode frequency, $\omega_{LO}(q \approx 0)$, of PbTe is well known from inelastic neutron scattering (Cochran et al., 1966; Alperin et al., 1972) and tunneling studies (Takasaki and Tanaka, 1977). A temperature-independent value of 114 cm$^{-1}$ is generally accepted. In alloy systems, a slight variation of $\omega_{LO}$ with composition was found (Takasaki and Tanaka, 1977).

Thus, to fit the reflectivity data in the neighborhood of the TO resonance we are left with two adjustable parameters: the soft-mode frequency $\omega_{TO}(q \to 0)$, and the phonon damping parameter $\Gamma$. No frequency dependence of $\Gamma$ is required for an optimum fit of the FIR spectra.

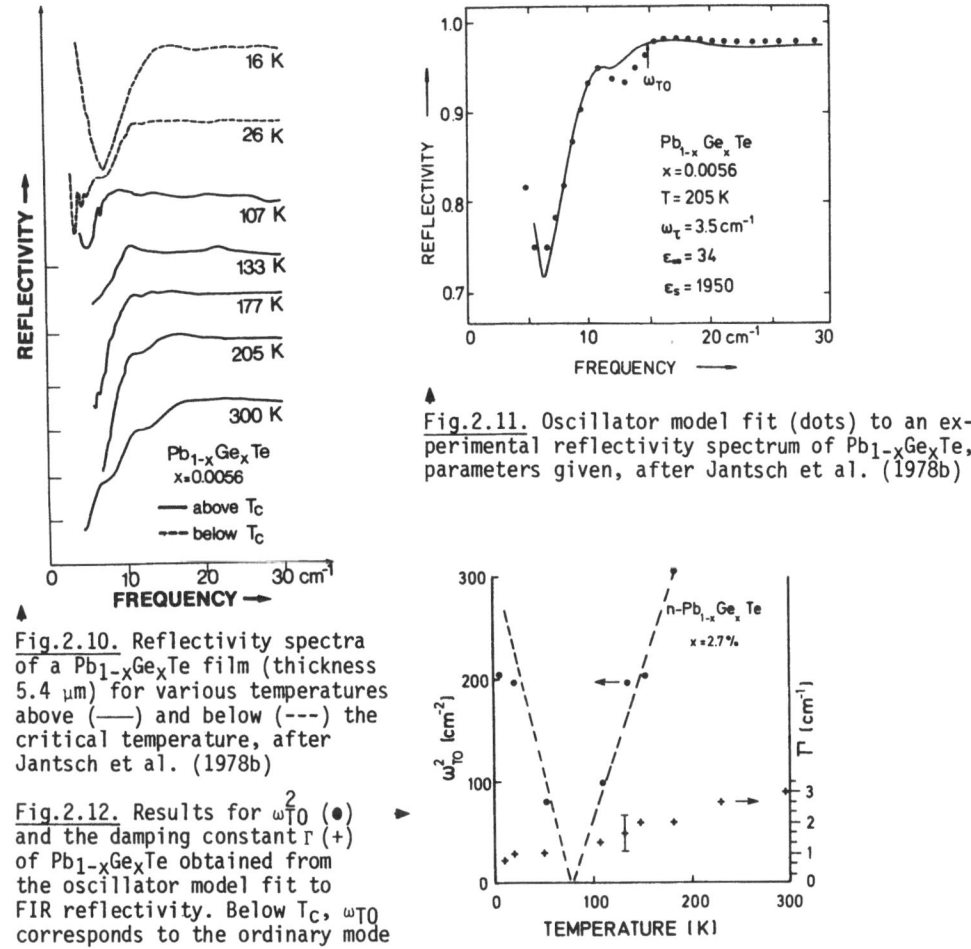

Fig.2.11. Oscillator model fit (dots) to an experimental reflectivity spectrum of $Pb_{1-x}Ge_xTe$, parameters given, after Jantsch et al. (1978b)

Fig.2.10. Reflectivity spectra of a $Pb_{1-x}Ge_xTe$ film (thickness 5.4 μm) for various temperatures above (——) and below (---) the critical temperature, after Jantsch et al. (1978b)

Fig.2.12. Results for $\omega_{TO}^2$ (●) and the damping constant $\Gamma$ (+) of $Pb_{1-x}Ge_xTe$ obtained from the oscillator model fit to FIR reflectivity. Below $T_c$, $\omega_{TO}$ corresponds to the ordinary mode

The FIR method has been used for investigations of $\omega_{TO}$ in PbTe (Buss and Kinch, 1973; Burkhard et al., 1976; Bauer et al., 1978) and $Pb_{1-x}Ge_xTe$ (Jantsch et al., 1978b, 1979). For GeTe and SnTe this method is less suitable, since samples of these compounds have carrier concentrations of the order of $10^{20}$ cm$^{-3}$, whereas PbTe and $Pb_{1-x}Ge_xTe$ samples with carrier concentrations of $10^{16} - 5 \cdot 10^{17}$ cm$^{-3}$ are available. For high carrier concentrations the negative contribution from free carriers, $Re\{\tilde{\varepsilon}_{FC}\}$, dominates the FIR optical properties and the dielectric anomaly at the TO resonance becomes too small for experimental observation.

Experimental results for the reflectivity of <111> oriented thin films of $Pb_{1-x}Ge_xTe$ are given in Fig.2.10 for various temperatures above and below the critical temperature of $T_c = 36$ K for this particular sample (Jantsch et al., 1978b). Above $T_c$, the onset of the drop in reflectivity is shifted towards lower frequencies with decreasing temperature. This behavior indicates softening of the TO mode. Below $T_c$, the soft mode is stabilized and the structure in the reflectivity extends to higher frequencies again. Figure 2.11 examplifies the oscillator model fit in this spectral regime. The solid line is experimental, the dots result from the oscillator model.

Typical results for the temperature dependence of the soft-mode frequency and its damping constant $\Gamma$, obtained from the FIR method, are given in Fig.2.12. The critical temperature $T_c$ is obtained from linear extrapolation of $\omega_{TO}^2 \rightarrow 0$. Above and below $T_c$, the curves rise linearly according to a Curie-Weiss law. A low-frequency limit for the determination of the soft-mode frequency of 10 cm$^{-1}$ is imposed experimentally, which corresponds to an inaccessible temperature interval of $|T - T_c| \lesssim 20$ K. In this critical regime the mode softening cannot be studied by FIR- or Raman-spectroscopy.

In the rhombohedral phase, splitting of the phonon modes is expected due to the lowering of crystal symmetry: the two low-frequency Raman lines have been interpreted accordingly (Fig.2.3). The structure observed in the FIR reflectivity ($\omega < 20$ cm$^{-1}$) yields a TO frequency comparable to that of the lower frequency Raman line, which was attributed to the ordinary $E_{TO}$ modes. No evidence was found for a second phonon mode in the low-frequency regime ($\omega < 60$ cm$^{-1}$). The rest of this section is devoted to a discussion of mode splitting due to the rhombohedral distortion.

In the rhombohedral phase, the 3-fold degeneracy of the cubic $F_{1u}$ mode is lifted by the lowering of symmetry at the phase transition. In the rhombohedral ($C_{3v}$) phase a doubly degenerate, "ordinary" E mode, and a single "extraordinary" $A_1$ mode exist. For the $A_1$ mode the two sublattices oscillate along the rhombohedral axis with respect to each other, the E modes correspond to a relative motion perpendicular to the c-axis. For the ordinary mode transverse solutions exist: the electric field $\underline{E}$ and the polarization $\underline{P}$ are perpendicular to both the c-axis and the wave-propagation direction $\underline{k}$. The E modes are isotropic: they do not show directional dispersion.

For the extraordinary wave, however, the polarization in general is not perpendicular to the c-axis or $\underline{k}$. Therefore the $A_1$ mode contains both transverse- and longitudinal-components. Its frequency depends on the angle $\theta$ between $\underline{k}$ and the c-axis.

The dielectric function $\tilde{\varepsilon}_E(\omega)$ of the E modes has the same form as that of the $F_{1u}$ modes (2.5) (Loudon, 1964; Claus et al., 1975):

$$\tilde{\varepsilon}_E(\omega) \equiv \tilde{\varepsilon}^{\perp}(\omega) = \varepsilon_{\infty}^{\perp} \frac{\omega_{E_{LO}}^2 - \omega^2 - i\omega\Gamma^{\perp}}{\omega_{E_{TO}}^2 - \omega - i\omega\Gamma^{\perp}} \quad . \tag{2.12}$$

For the extraordinary mode the following expression is obtained:

$$\tilde{\varepsilon}_{A_1}(\omega,\theta) = \frac{\tilde{\varepsilon}^{\perp}(\omega) \cdot \tilde{\varepsilon}''(\omega)}{\tilde{\varepsilon}^{\perp}(\omega)\sin^2\theta + \tilde{\varepsilon}''(\omega)\cos^2\theta} \quad , \tag{2.13}$$

where

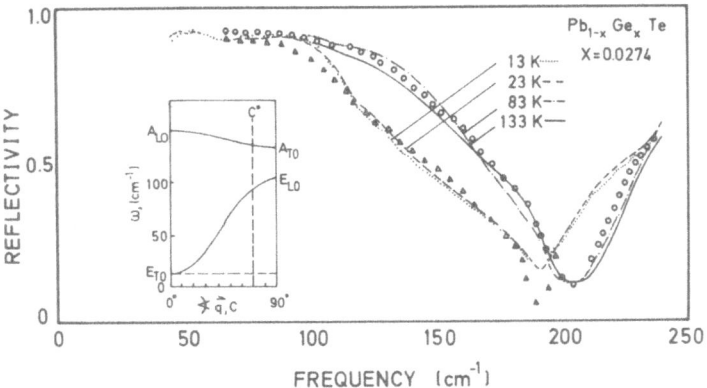

Fig.2.13. Reflectivity of a $Pb_{1-x}Ge_xTe$ film grown on a $BaF_2$-(111)-surface for temperatures above and below the critical temperature of $T_c = 73$ K. The drop in reflectivity above 100 cm$^{-1}$ is due to the combined plasmon-phonon edge. Experimental data
(lines) are compared to results from a two-phonon oscillator model fit (circles
and triangles). The inset shows the directional dispersion of the $A_1$- and E-phonons,
corresponding to the best-fit parameters obtained (triangles), from Jantsch et al.
(1979)

$$\tilde{\varepsilon}''(\omega) = \varepsilon_\infty'' \frac{\omega_{A_{1LO}}^2 - \omega^2 - i\omega\Gamma''}{\omega_{A_{1TO}}^2 - \omega^2 - i\omega\Gamma''} . \tag{2.14}$$

In (2.12-14) quantities related to a polarization parallel and perpendicular to the
c-axis are designated by the superscripts " and $\perp$, respectively.

The total phonon dielectric function including the high-frequency dielectric
constants is given by the sum of (2.12,13). For k‖c ($\theta = 0$), (2.13) reduces to
(2.12). In this case, the low-frequency dielectric anomaly is caused only by the
$E_{TO}$ mode. The reflectivity data from [111] oriented samples (Fig.2.10) have been
interpreted accordingly. However, in the neighborhood of the plasma edge, the spectra given in Fig.2.13 show an anomaly (Jantsch et al., 1979): below $T_c$, the combined plasmon-phonon edge shows additional structure. If the rhombohedral distortion occurs only in that [111] direction, which is perpendicular to the sample surface, only the E modes are observable. The additional structure can be explained
if we assume that ferroelectric domains with their c-axis along the other, equivalent cubic <111> directions are also present in the rhombohedral phase.

Ferroelectric domains in Group IV tellurides have been observed by electron
microscopy (Snykers et al., 1972; Jantsch et al., 1981b). The interpretation of
cyclotron resonance spectra (Lewis et al., 1980) gives further evidence for the
existence of ferroelectric domains in $Pb_{1-x}Ge_xTe$. For [111] oriented samples, two
types of domains exist: a-domains with their c-axes perpendicular to the surface,
and b-domains, whose c-axes include an angle of about 70° with the c-axis of the
a-domains. According to Jantsch et al. (1981b), the domain size in $Pb_{1-x}Ge_xTe$ is

19

in the micron range, which is small compared to the wavelength of FIR radiation. Therefore the optical properties can be described by an average dielectric function containing weighted contributions from both $\tilde{\varepsilon}_E(\omega)$ and $\tilde{\varepsilon}_{A_1}(\omega, \theta = 70°)$. Assuming equal populations of the four possible domain orientations, the spectra given in Fig.2.13 have also been fitted for $T < T_c$. Results from the oscillator model fit are given in Fig.2.13 together with the corresponding directional dispersion. The level ordering obtained, $\omega_{A_{TO}} > \omega_{E_{LO}}$, does not agree with that from an interpretation of Raman data (Sect.2.3). At low temperatures, samples of comparable composition and critical temperature show Raman lines at 30-50 cm$^{-1}$, which have been attributed to the $A_1$ mode due to their directional dispersion. An extensive search in this spectral regime using FIR spectroscopy failed to show any evidence for an extraordinary mode at a frequency different from the $E_{TO}$ mode.

The level ordering depends on the relative magnitude of the anisotropy of the force constants due to the rhombohedral distortion and, on the other hand, the electrostatic ones, which are responsible for the difference between longitudinal and transverse modes (Loudon, 1964). In the present case, the rhombohedral distortion is rather small. Therefore, we expect the electrostatic forces to be larger than their anisotropy and hence a level ordering according to: $|\omega_{A_{1TO}} - \omega_{E_{TO}}| \ll (\omega_{A_{1LO}} - \omega_{A_{1TO}})$ and $(\omega_{E_{LO}} - \omega_{E_{TO}})$, in agreement with the Raman data.

Alternative interpretations of the additional structure in the neighborhood of the plasma edge below $T_c$ have been discussed recently (Jantsch et al., 1981b). In the rhombohedral phase the plasma effective mass depends on the angle $\theta$ as well. Thus, the free-carrier contribution to a- and b-domains in the dielectric function will be different. Using the effective mass tensor obtained from cyclotron resonance investigations, the reflectivity spectra can be explained quantitatively with a corresponding model by assuming a negligible E-$A_1$ phonon mode splitting (Jantsch et al., 1981b). The E-$A_1$ splitting is not resolvable in the FIR spectra, although the damping parameter $\Gamma$ (Fig.2.12) is much smaller than the line width observed in Raman experiments, since the latter is strongly enhanced due to the extremely small penetration depth (Sect.3.2).

## 3.  Experimental Determination of the Static Dielectric Constant

### 3.1  Differential Capacitance Measurements

The best and purest samples of IV-VI compounds exhibit concentrations of $10^{16}$-$10^{20}$ cm$^{-3}$ of highly mobile carriers due to intrinsic defects. The standard methods to determine the static dielectric constant, which are based on measurements of the complex low-frequency impedance of capacitors filled with the sample dielectric, are not applicable because of the high conductivity of these materials. The problems resulting from the dominating real part of the conductivity are avoided by measuring the dielectric constant in the space charge region of p-n junctions, Schottky barriers or MIS structures.

20

Investigations on junctions were first reported by Kanai and Shohno (1963) for PbTe. The capacitance C of an abrupt junction with an applied bias V is given by (Sze, 1969)

$$C/A = \left[ \frac{\varepsilon_s Ne}{8\pi(V + V_{bi})} \right]^{1/2} , \qquad (3.1)$$

where A is the junction area, $\varepsilon_s$ the static dielectric constant, (Ne) the space charge density and $V_{bi}$ the built-in voltage of the diode. From (3.1) we obtain

$$\frac{\partial}{\partial V} \left( \frac{1}{C^2} \right) = \frac{8\pi}{\varepsilon_s NeA^2} . \qquad (3.2)$$

Thus, linear curves are obtained by plotting $C^{-2}$ as a function of the bias voltage. The slope given in (3.2) depends on $\varepsilon_s$ and N. The space-charge density of homogeneous samples equals the bulk carrier concentration determined from the d.c. Hall effect times the elementary charge. Thus, the static dielectric constant is obtained from the slope of $C^{-2}$-V curves using (3.2). A detailed investigation of Schottky barriers on PbTe (Walpole and Nill, 1971) and $Pb_{1-x}Sn_xTe$ (Nill et al., 1971) has shown that (3.1) describes experimental C-V characteristics even when inversion layers exist at the surface. The name "differential capacitance method" derives from (3.2), although the junction capacitance is measured by conventional techniques using radio frequency bridges or lock-in amplifiers without differentiation.

Kanai and Shono (1963) investigated n- and p-type samples of PbTe with carrier concentrations between $1.5 \cdot 10^{17}$ and $8 \cdot 10^{18}$ cm$^{-3}$. Abrupt junctions were obtained by alloying In to p-type and Ag-Te to n-type PbTe. For temperatures between 4.2 K and 130 K a constant value of $\varepsilon_s$ = 400 was found, independent of carrier type and concentration.

For paraelectric materials we expect a linear dependence of $\varepsilon_s^{-1}$ on temperature, according to the Curie-Weiss law. This behavior was first observed by Bate et al. (1970) on diffused p-n junctions of PbTe. For linearly graded junctions (Sze, 1969)

$$C/A = \left[ \frac{\varepsilon_s^2(\partial N/\partial x)e}{36\pi(V + V_{bi})} \right]^{1/3} , \qquad (3.3)$$

is obtained. Here $\partial N/\partial x$ denotes the constant doping gradient across the junction. A linear dependence of $C^{-3}$ on bias voltage was observed on PbTe and $Pb_{0.83}Sn_{0.17}Te$ diodes (Bate et al., 1970). Only at low temperatures deviations from linearity were found on PbTe and $Pb_{1-x}Sn_xTe$ diodes of different doping gradients. The deviations became apparent essentially at the same threshold polarization of about $10^{-6}$ C/cm$^2$. The maximum polarization $P_{max}^{lg}$ within the linear graded junction was estimated from

$$P_{max}^{lg} \simeq \frac{3}{2} \frac{C}{A} (V + V_{bi}) . \qquad (3.4)$$

21

Equation (3.4) is obtained from the field-strength maximum in the space charge region of a graded junction (Sze, 1969):

$$E(x) = -e \frac{\partial N}{\partial x} \frac{(w/2)^2 - x^2}{2\varepsilon_s} \quad . \tag{3.5}$$

Here w stands for the depletion layer width:

$$w = \frac{A\varepsilon_s}{C} \quad . \tag{3.6}$$

Combining (3.5) for $x = 0$ with (3.6,3) and $P \simeq \varepsilon_s E$ for $\varepsilon_s \gg 1$, then (3.4) is obtained. The nonlinearities of $C^{-3}$ versus V at low temperatures and high reverse bias V were attributed to a nonlinear dependence of the polarization P on the electric field (Cochran, 1960):

$$E = c_1(T - T_c)P + c_3 P^3 + c_5 P^5 + \ldots \quad . \tag{3.7}$$

Bate's et al. (1970) results for $\varepsilon_s$ are qualitatively different from those of Kanai and Shono (1963), who did not observe the temperature dependence of $\varepsilon_s$. This discrepancy was attributed to a maximum polarization of $40 \cdot 10^{-6}$ C/cm$^2$, estimated from the data of Kanai and Shono. This value seems to indicate that a polarization well in excess of the quoted threshold value of $1 \cdot 10^{-6}$ C/cm$^2$ may occur in C-V experiments on abrupt junctions. In this situation, nonlinear terms in (3.7) would dominate and $\varepsilon_s$ and hence $\partial C^{-2}/\partial V$ in (3.1) would depend on the applied bias voltage, in contrast to the experimental results of Kanai and Shono.

Unfortunately Bate et al. did not explain how the maximum polarization was estimated for the abrupt junction. From a treatment analogous to that described above for graded junctions, we obtain for abrupt junctions a maximum polarization of

$$P_{max}^a \simeq 2(V + V_{bi}) \frac{C}{A} \quad . \tag{3.8}$$

From this expression and the experimental data of Kanai and Shono a maximum polarization of $1.5 \cdot 10^{-6}$ C/cm$^2$ is estimated, comparable to the value obtained for graded junctions. It should be pointed out that the deviations observed by Bate et al. (1970) on linear graded junctions at low temperatures can be also explained in terms of doping profile nonlinearities: the width of the space-charge region w of linear graded junction is obtained from (Sze, 1969):

$$w^3 = \frac{12\varepsilon_s(V + V_{bi})}{e(\partial N/\partial x)} \quad . \tag{3.9}$$

For a specific bias voltage V, the depletion width increases with increasing dielectric constant. Since the dielectric constant increases with decreasing temperature in paraelectric PbTe, w increases with decreasing temperature. An evaluation of Bate's et al. (1970) experimental data shows that the threshold voltage observed at 4.2 K corresponds to a depletion width unobtainable at the next higher temperature of 77 K within the investigated voltage range. The positive deviations of the $C^{-3}$-V curves observed at 4.2 K indicate a flattening of the doping gradient, which is expected for large distances away from the junction.

Antcliffe et al. (1973) reported capacitance measurements on $Pb_{1-x}Ge_xTe$ junction diodes at various fixed dc biases as a function of temperature. At the phase transition temperature, a maximum of the diode capacitance was observed, indicating the highest lattice polarizability at this point. The transition temperature derived from this method is in good agreement with values obtained by Hohnke et al. (1972) from X-ray analysis. Antcliffe et al. also observed additional structure in the temperature dependence of the diode capacitance and a tentative interpretation was given in terms of an electric-field-induced phase transition at temperatures well above the phase transition temperature. However, there is no other indication in the literature of the appearance of this effect, and other interpretations of the reported anomalies are possible in the absence of a detailed investigation of the diode characteristics.

An influence of lattice defects on the dielectric behavior of PbTe (Jantsch and Lopez-Otero, 1976) and $Pb_{1-x}Ge_xTe$ (Jantsch et al., 1981a,1982) was postulated from investigations of the capacitance of Schottky diodes. Differential capacitance measurements on Schottky diodes allow a determination of absolute values of $\epsilon_s$ in contrast to measurements on graded junctions, since in the latter case the doping gradient is experimentally inaccessible.

In Fig.3.1 results for the inverse quasi-static dielectric constant $\epsilon_s^{-1}$ are given as a function of temperature for two p-type samples of PbTe with different carrier concentrations. Here $\epsilon_s$ is designated as quasi-static, since measurements are performed in the MHz regime. True static measurements are not possible because of leakage currents in Schottky diodes on IV-VI compounds. The maximum polarization involved here was estimated from (3.8) to be less than $10^{-6}$ $C/cm^2$. At high temperatures, $\epsilon_s$ varies linearly with temperature according to the Curie-Weiss law, whereas at low temperatures the inverse dielectric constant saturates due to zero-point fluctuations (Barrett, 1952; Müller and Burkard, 1979). The inverse dielectric constant depends strongly on the carrier concentration of the sample. In the C-V method, however, the dielectric constant is determined in a space-charge region in the absence of free carriers. In undoped PbTe, free carriers originate from doubly ionized Pb and Te vacancies (Pratt, 1973,1974; Heinrich et al., 1975,1980). Therefore the apparent influence of the carrier concentration on $\epsilon_s$ is attributed to vacancies rather than to the free carriers. Additional evidence for this conclusion is derived from a variation of the crystal growth conditions (Jantsch and Lopez-Otero, 1976) and recently also from investigations of electron-irradiated $Pb_{1-x}Sn_xTe$ samples (Suski et al., 1982).

Samples of equal carrier concentration grown at different temperatures also exhibit different dielectric behavior. The carrier concentration of compensated samples is given by

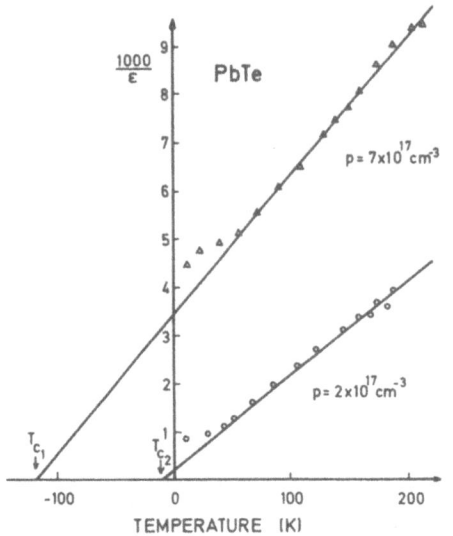

**Fig.3.1.** Temperature dependence of the inverse quasi-static dielectric constant of two p-PbTe samples of different hole concentration, after Jantsch and Lopez-Otero (1976)

$$N = 2[V_{Te}] - 2[V_{Pb}] + [I_{Pb}] \quad , \tag{3.10}$$

where $[V_{Pb}]$ and $[V_{Te}]$ stand for the concentration of Pb and Te vacancies, respectively, and $[I_{Pb}]$, for the concentration of interstitial Pb atoms (Heinrich, 1980). Obviously, the total number of point defects and hence the dielectric properties depend on the growth conditions, whereas because of compensation the carrier concentration need not change. Unfortunately, the usual methods for determining the compensation or the total number of defects are not applicable for IV-VI compounds because of the special nature of the vacancy states. Ionized impurity scattering is very weak because of the high dielectric constant (Allgaier and Scanlon, 1958) and/or the short-range nature of the vacancy potential (Pratt, 1974). Defect spectroscopy methods do not work either, since the native donors and acceptors in PbTe form resonant states (Pratt, 1974; Gresslehner et al., 1978). Therefore, at present a quantitative evaluation of the influence of defects is not possible.

Differential capacitance measurements have also been performed on $Pb_{1-x}Ge_xTe$ (Grishechkina et al., 1978; Jantsch et al., 1981a,1982). The alloy system $Pb_{1-x}Ge_xTe$ undergoes a phase transition for $x > 0.005$ (Jantsch et al., 1979). Results for the temperature dependence of $\varepsilon_s^{-1}$ obtained from differential capacitance measurements on Schottky barriers are given in Fig.3.2 (Jantsch et al., 1981a). For comparison, data for the static dielectric constant, $\varepsilon_0$, derived by the FIR method (Sect.2.4) via the Lyddane-Sachs-Teller relation (1.1) are also given. The results obtained with the quasi-static method show substantial deviations from those obtained from the phonon data. The inverse quasi-static dielectric constant $\varepsilon_s^{-1}$ exhibits a minimum at a temperature, which coincides with the critical temperature obtained with the FIR methods. However, extrapolations of the more or less linear parts above and below $T_c$ intersect at $\varepsilon_s^{-1} \neq 0$. The $\varepsilon_s^{-1}$ curves are shifted towards higher

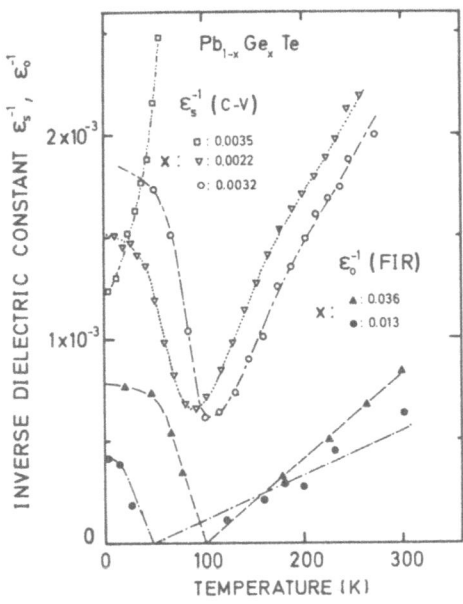

INVERSE DIELECTRIC CONSTANT $\varepsilon_s^{-1}$, $\varepsilon_o^{-1}$

$Pb_{1-x}Ge_xTe$

$\varepsilon_s^{-1}$ (C-V)

□ : 0.0035
X :  ▽ : 0.0022
     ○ : 0.0032

$\varepsilon_o^{-1}$ (FIR)

X :  ▲ : 0.036
     ● : 0.013

TEMPERATURE (K)

**Fig.3.2.** Comparison of results for the inverse quasi-static dielectric constant $\varepsilon_s^{-1}$ obtained from C-V measurements, and the static dielectric constant $\varepsilon_0^{-1}$ calculated from FIR phonon data using the LST relation, after Jantsch et al. (1981a)

values with respect to the $\varepsilon_0^{-1}$ data, which were obtained from the phonon frequencies. This discrepancy indicates that additional dispersion occurs in the frequency range between the Reststrahlenband in the THz range and the low-frequency MHz regime, where $\varepsilon_s$ is determined. The same effect obviously causes the variation of $\varepsilon_s^{-1}$ of PbTe with defect concentration. Since PbTe is paraelectric, no minimum of $\varepsilon_s^{-1}$ occurs at finite temperatures and a negative Curie temperature is obtained by extrapolating $\varepsilon_s^{-1} \rightarrow 0$. This Curie temperature is just a parameter which describes the dielectric properties rather than the crystal stability. The latter is characterized by a critical temperature derived from phonon data (Fig.5.3). In Fig.3.1 the apparent variation of $T_c$ with defect concentration is mainly caused by the additional polarizability in the low-frequency range below the Reststrahlen regime (Sect.5.3).

### 3.2 Microwave Techniques

A different approach to the experimental determination of the static dielectric constant of IV-VI compounds is based on investigations of the optical properties in the microwave regime in the presence of high magnetic fields. Without magnetic field, electromagnetic waves with frequencies below the Reststrahlen-plasma regime cannot propagate into the material, since the real part of the dielectric function (2.2) is negative. In the presence of a magnetic field B the free-carrier part of the dielectric function, $\tilde{\chi}_{FC}$, becomes a second-rank tensor even for materials with isotropic band structure. A general introduction to the field of magnetoplasma effects in semiconductors was given by Palik and Furdyna (1970).

In the cubic phase of PbTe, SnTe and GeTe, the constant energy surfaces of the valence- and conduction band are described by ellipsoids oriented along the <111> directions. A discussion including all possible configurations of the magnetic field B and the direction of light propagation relative to the crystal axes would be tedious. A number of configurations has been investigated by Stiles et al. (1962) Nii (1964), Perkowitz (1969), Takano et al. (1974) and Foley and Langenberg (1977).

In the case of highest symmetry, B||<100>, and Faraday configuration k||B, the two normal magnetoplasma modes are circularly polarized waves. The dispersion of the magnetoplasma waves is described by the dielectric function

$$k_\pm^2 = \left(\frac{\omega}{c}\right)^2 (\tilde{\varepsilon}_{xx} \pm i\tilde{\varepsilon}_{xy}) \equiv \left(\frac{\omega}{c}\right)^2 \tilde{\varepsilon}_\pm(\omega,B) \tag{3.11}$$

with

$$\tilde{\varepsilon}_\pm(\omega,B) = \varepsilon_\infty + \tilde{x}_{ph}(\omega) + \tilde{x}_{FC}^\pm(\omega,B) \quad, \tag{3.12}$$

where the plus and minus signs correspond to right- and left-circular polarized waves. In the local approximation, $\tilde{x}_{FC}^\pm$ is given by (Wallace, 1965)

$$\tilde{x}_{FC}^\pm = \frac{\omega_p^2}{\omega} \frac{\omega_3 \pm (\omega + i\omega_\tau)}{(\omega + i\omega_\tau)^2 - \omega_c^2} \quad, \tag{3.13}$$

where

$$\omega_p^2 = \frac{4\pi N e^2}{m_t^*} \frac{2K + 1}{3K} \quad, \quad K = \frac{m_\ell^*}{m_t^*}$$

$$\omega_c^2 = \frac{e B^2}{m_t^*} \frac{K + 2}{3K} \quad, \quad \omega_3 = \frac{e B}{m_t^*} \frac{K + 2}{2K + 1} \quad .$$

Here $m_\ell^*$ and $m_t^*$ stand for the longitudinal- and transverse-effective mass, respectively, and N is the free-carrier concentration.

The main structures in the optical properties result from two types of singularities in the dielectric function: poles, which are caused by cyclotron- or TO-resonances, and zeros associated with longitudinal excitations. Points where Re{ε} = 0 are referred to as "dielectric anomalies."

A discussion of optical properties is simplified by considering contour maps of the zeros and poles of the dielectric function in the $\omega$ versus $\omega_c$ plane in the limit of negligible damping $\Gamma$ and $\omega_\tau$ (Palik and Furdyna, 1970). A contour map derived from (3.12,13) is given in Fig.3.3. The solid lines indicate the zeros and the dashed lines the poles of $\tilde{\varepsilon}_\pm$. In the shaded regions the real part of $\tilde{\varepsilon}_\pm$ becomes negative and wave propagation into the material is not possible since the refractive index $n^\pm = \sqrt{\tilde{\varepsilon}_\pm}$ becomes imaginary for Re{$\tilde{\varepsilon}_\pm$} < 0. In this regime total reflection occurs. For B = 0 the situation is exactly that encountered in the FIR reflectivity (Sect. 2.4). The dielectric function is positive only above the plasma edge and in a narrow

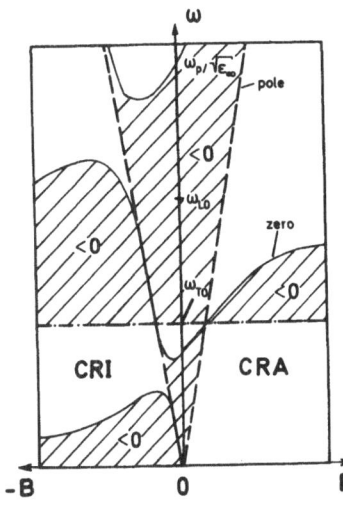

**Fig.3.3.** Contour map of the dielectric function for $B\|<\bar{1}00>$ and Faraday geometry. The solid lines indicate zeros of the dielectric function; the dashed lines represent the poles due to the cyclotron resonance; and the dash-dotted line the TO-phonon resonance. The magnetic field direction is chosen positive for the cyclotron resonance active (CRA) configuration, the negative $\underline{B}$ direction corresponds to the cyclotron resonance inactive (CRI) situation, after Foley and Langenberg (1977)

region just below the TO resonance. In the presence of a sufficiently high magnetic field, however, the dielectric function becomes positive again in some areas of the contour map even for $\omega \ll \omega_{TO}$. The magnetic field required for $\tilde{\epsilon}_\pm > 0$ also depends on the direction of the magnetic field. (The positive direction of B is defined as that for which the cyclotron orbit has the same sense of rotation as the circular polarized incident wave.) In the transmission or reflectivity a dielectric anomaly causes a sharp increase or drop, when the magnetic field is scanned across one of the values represented by solid lines in Fig.3.3. The position of dielectric anomalies depends on the lattice contribution $\epsilon_\infty + \tilde{\chi}_{ph}(\omega)$, as can be seen by setting $\tilde{\epsilon}_\pm(\omega) = 0$ in (3.12). Thus, the lattice contribution at a frequency $\omega$ can be determined from the position of the dielectric anomalies if the other parameters, namely the carrier concentration N and the effective masses, are known (Perkowitz, 1969; Foley and Langenberg, 1977). The effective mass parameters can be obtained from cyclotron resonance in different geometries. Since the cyclotron resonance position is given by a pole of $\tilde{\chi}_{FC}^\pm(\omega)$ it does not depend on the lattice contribution. The carrier concentration is derived either from d.c. Hall experiments or from magnetooptical experiments in the so-called helicon regime, or from a self-consistent method which involves measurements at different frequencies (Foley and Langenberg, 1977).

Alternatively, the lattice contribution to the dielectric function at a frequency $\omega < \omega_{TO}$ and carrier concentration N can be determined from the refractive index as a function of the magnetic field above the cyclotron resonance field. The refractive index is determined from Fabry-Perot interference effects in plane parallel slabs, which cause periodic changes of the reflectivity versus $B^{-1}$ (Takano et al., 1974; Kawamura et al., 1975).

The static dielectric constant is evaluated from the lattice contribution $\epsilon_\infty + \tilde{\chi}_{ph}(\omega)$, making use of (2.5), which includes the mode frequencies $\omega_{TO}$, $\omega_{LO}$ and their damping parameter. Assuming values from the literature for $\omega_{LO}$ and $\varepsilon_\infty$, and weak damping, the TO-mode frequency can be obtained. From $\omega_{TO}$, the static dielectric constant $\varepsilon_0$ can be obtained via the Lyddane-Sachs-Teller relation. Most experiments are performed at microwave frequencies ($\omega \ll \omega_{TO}$) where $\varepsilon_0 \approx \varepsilon_\infty + \chi_{ph}(\omega)$. Thus, the measured lattice part of the dielectric function is assumed to be equal to the phononic static dielectric constant $\varepsilon_0$.

A summary of experimental results for the static dielectric constant of Group IV tellurides at low temperatures obtained from the magnetoplasma method is given in Table 3.1.

Table 3.1. Static dielectric constant of $Pb_{1-x}Ge_xTe$ obtained from magnetoplasma experiments

| Compo-sition $x$ | Tempera-ture [K] | Carrier conc. $10^{17}$ cm$^{-3}$ (<0 for n-type) | Freq. [GHz] | $\varepsilon_0$ at 4.2 K | Reference |
|---|---|---|---|---|---|
| 0 | 4.2 | 5 | 70 | $10^4$ | Sawada et al. (1965) |
| 0 | 0.2 | -8.1 | 70 | $10^4$ | Perkowitz (1969) |
| 0-0.293 | 1.6,4.2 | -1.24 to -11.4 | 50 | 660 -10800 | Takano et al. (1974) Kawamura et al. (1975) |
| 0-0.4 | 4.2-100 | -1 to -20 | 35,50 | 1500-60000 | Nishi et al. (1980) |
| 0,0.3 | 4.2 | -2,+1.3 | 580-965 | 1210,13700 | Ichiguchi et al. (1980) |
| 0 | 4.2-80 | 0.54-5.4,-0.8-5 | 70,138,335 | 3000 | Foley and Langenberg (1977) |

A systematic dependence of $\varepsilon_s$ of $Pb_{1-x}Sn_xTe$ on the effective minimum energy gap was reported by Takano et al. (1974). The effective minimum energy gap is obtained by taking into account the enlargement of the gap due to population of states close to the band edge in the presence of a degenerate electron gas. The influence of the effective energy gap on the dielectric properties was attributed to interband electron-phonon interaction, according to the model by Kristoffel and Konsin (1968). However, recent investigations with the same method failed to show any influence of the carrier concentration in the range of $1-20 \cdot 10^{17}$ cm$^{-2}$ (Nishi et al., 1980), or of the zero-gap transition, which occurs in $Pb_{1-x}Sn_xTe$ at $x \approx 0.40$ at 70 K.

Microwave impedance-measurement techniques for samples with low carrier concentration and not too high mobilities can also be used to determine the static dielectric constant even without a magnetic field (Foley and Langenberg, 1976; Lehmann et al., 1981). In this method the static dielectric constant is obtained from the complex conductivity. The results by Lehmann et al. (1981) again indicate that the static dielectric constant does not depend systematically on the effective energy gap.

## 4. Effects Related to the Phase Transition

Besides their dielectric and optical properties, several other effects related to the structural phase transition have been investigated in (PbSnGe)Te compounds. The phase transition manifests itself in an anomaly of the temperature dependence of some measured quantity in the neighborhood of the critical temperature. In the following, some of these effects are reviewed.

### 4.1  Changes of Band Structure and Related Phenomena

In the cubic phase, the lowest conduction band minima and the highest valence band maxima of $Pb_{1-x}Ge_xTe$ and $Pb_{1-x}Sn_xTe$ are located at the L points of the Brillouin zone. The rhombohedral distortion below $T_c$ consists in a relative displacement of the two sublattices along a [111] direction, which becomes the rhombohedral c-axis. Due to this distortion, the four <111> directions of the cubic phase are no longer equivalent: the valley oriented along the c-axis (T valley) is different in energy and curvature compared to the remaining three L valleys, which are still equivalent in the rhombohedral phase (Lewis et al., 1980).

Evidence for valley splitting was found from cyclotron absorption in $Pb_{1-x}Ge_xTe$ (Lewis et al., 1980). Experiments were performed on epitaxial (111) oriented $Pb_{1-x}Ge_xTe$ films in Faraday geometry using FIR lasers. In the cubic phase, two absorption peaks were observed as a function of the magnetic field, one due to the transverse cyclotron resonance of the valley oriented along the surface normal, the other caused by the tilted orbit resonance of the remaining three L valleys. The axes of the corresponding ellipsoidal constant-energy surfaces are tilted by $70^\circ$ with respect to the surface normal. Below $T_c$, four absorption peaks were observed, indicating that two types of ferroelectric domains must exist in the sample, one with its c-axis perpendicular to the sample surface ("a" domains) and the other whose c-axis is oriented along one of the remaining oblique <111> directions ("b" domains). From the position of the cyclotron resonance peaks, effective band-edge mass values were obtained (Lewis et al., 1980) taking into account the non-parabolicity of the bands. From the relative amplitudes of the absorption peaks the population ratio of a- and b-domains was calculated. For films grown on $BaF_2$, it was found that b-domains build up preferentially due to the strain in the sample plain, which results from different expansion coefficients of film and substrate.

Anomalies of the refractive index and the fundamental absorption edge of $Pb_{1-x}Ge_xTe$ have also been explained in terms of the L-T valley splitting (Jantsch, 1980). In the rhombohedral phase, the absorption edge of $Pb_{1-x}Ge_xTe$ (x > 0.09) exhibits a tail extending into the forbidden energy gap. This additional absorption is caused by indirect L-T transitions. The high-frequency dielectric constant of IV-VI compounds contains a relatively large contribution from interband transitions close to the fundamental energy gap (Cardona, 1968). Thus, the birefringence observed in $Pb_{1-x}Ge_xTe$

(x > 0.09) can be attributed to the L-T level splitting of the conduction and valence bands (Jantsch, 1980). Correlation between the birefringence and L-T valley splitting is also evident from the observation that the birefringence vanishes in samples with high carrier concentrations. The rhombohedral distortion also causes a change in the temperature coefficient of the direct energy gap and the associated refractive index (Murase et al., 1979; Jantsch, 1980).

## 4.2 Resistance Anomaly

The resistivity versus temperature of SnTe (Kobayashi et al., 1975,1976; Grassie et al., 1979) and $Pb_{1-x}Sn_xTe$ and $Pb_{1-x}Ge_xTe$ (Takaoka and Murase, 1979; Unterleitner, 1977) exhibits an anomalous increase of a few percent in the vicinity of the ferroelectric transition temperature obtained from other, more direct methods. This anomaly was interpreted by assuming that scattering of free carriers by phonons and other mechanisms causes a smooth background while the resistivity increment observed at the critical temperature is attributed to the carrier-soft-TO-phonon interaction (Kobayashi et al., 1975; Katayama, 1976). This interaction becomes particularly strong for $T \approx T_c$ due to the increase in phonon population as $\omega_{TO} \to 0$. Katayama and Mills (1980) treated this problem more thoroughly by also taking into account the anomalous scattering by the acoustic and LO phonons.

The influence of the free carrier concentration on the magnitude of the resistanc anomaly has been investigated for $Pb_{1-x}Sn_xTe$ (Murase, 1981; Murase and Nishi, 1982). Except for SnTe, which exhibits a quite different trend, experimental data for the anomalous increases in resistivity, $\Delta\rho(T_c)$, are consistent with

$$\Delta\rho(T_c) \propto T_c \, p^{-2/3} \quad,$$

where p is the hole concentration. A corresponding expression is obtained from the interband coupling model by assuming

$$(E_G + 2E_F)^2 = const. \quad,$$

which was not, however, very well satisfied for the samples under consideration.

A simple kink-shaped resistivity anomaly is observed only on "good samples". Different shapes of the resistivity anomaly have been found especially on $Pb_{1-x}Ge_xTe$ samples of low carrier concentration (Fig.4.1) (Unterleitner, 1977). Variation from sample to sample may be associated with the formation of ferroelectric domains in the rhombohedral phase and additional scattering by domain walls.

The resistance anomaly can be used as a simple tool to determine the critical temperature of (Pb,Sn,Ge)Te. The influence of the free carrier concentration (Kobayashi, 1975), hydrostatic pressure (Suski et al., 1979,1981), and high magnetic fields (Takaoka and Murase, 1979) on the critical temperature has been investigated using the resistance anomaly.

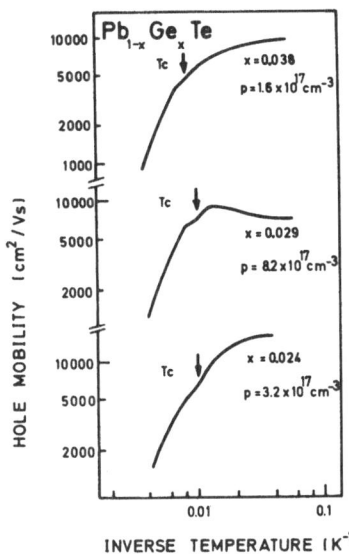

Fig.4.1. Hole mobility of $Pb_{1-x}Ge_xTe$ films on $BaF_2$ substrates versus inverse temperature. The arrows indicate the critical temperatures, after Unterleitner (1977)

### 4.3 Acoustic and Specific-Heat Anomalies

Investigations of ultrasonic wave velocities and attenuation measurements on $Sn_{1-y}Ge_yTe$ have been reported by Seddon et al. (1975,1976) and by Rehwald and Lang (1975) and on $Pb_{1-x}Ge_xTe$ by Sugai et al. (1979). In the vicinity of the phase transition the ultrasonic velocities show a steplike change accompanied by an attenuation peak. The results indicate that the phase transition in the alloy range considered is either continuous or only very weakly discontinuous. The acoustic anomaly is attributed to coupling of the acoustic branches at $q \approx 0$ to the soft TO mode. The anomalous specific heat observed in $Sn_{1-y}Ge_yTe$ was shown to be proportional to the anomalous elastic constant (Hatta and Rehwald, 1977).

The specific heat of $Sn_{1-y}Ge_yTe$ (Hatta and Rehwald, 1977), SnTe (Hatta and Kobayashi, 1977) and $Pb_{1-x}Ge_xTe$ (Sugimoto et al., 1981) also exhibits an anomaly at the critical temperature. According to Hatta and Rehwald (1977) and Sugimoto et al. (1981), this anomaly can be decomposed into two parts: (i) steplike behavior at $T_c$ and (ii) a fluctuation-dominated critical contribution around $T_c$. The steplike contribution

(i) can be explained by Landau theory. From the variation of the jump in specific heat with composition (x or y), indications for a cross-over from a second-order to a first-order phase transition were derived. A tricritical point was thus predicted for x or y ~ 0.6-0.7 (Sugimoto et al., 1981). From X-ray diffraction (Fig.1.2), a tricritical behavior has actually been observed at somewhat lower GeTe content (y ~ 0.28) (Clarke, 1978). The critical contribution

(ii) was found to fit a logarithmic law as predicted for uniaxial ferroelectrics (Larkin and Khmel'nitskii, 1969).

31

## 5. Results and Discussion

### 5.1 Temperature Dependence of the Soft Mode and Phase Transition

According to the soft-mode concept (Cochran, 1960; Anderson, 1960) a transverse optical mode frequency goes to zero at the critical temperature $T_c$ and the crystal structure changes. This correspondence can be examined experimentally by comparing results for $T_c$ obtained from crystal structure investigations with phonon data, which are available for temperatures not too close to $T_c$. In Fig.5.1, FIR results for the temperature dependence of $\omega_{TO}^2$ are given for $Pb_{1-x}Ge_xTe$ samples of different composition ($0 < x < 0.036$) (Jantsch et al., 1979). At high temperatures, $\omega_{TO}^2$ varies linearly according to the Curie-Weiss law:

$$\omega_{TO}^2 \sim \epsilon_s^{-1} = C_0(T - T_c)^\gamma , \tag{5.1}$$

where $\gamma$ is close to one. On approaching the critical temperature, $\omega_{TO}$ decreases. With both FIR spectroscopy and Raman scattering a low-frequency limit of about $10\ cm^{-1}$ for observing the soft mode is imposed experimentally. The critical temperature is therefore obtained by extrapolating the more or less linear dependence according to (5.1). Below $T_c$, the E-mode frequency observed in FIR experiments stabilizes again (Fig.5.1).

With increasing GeTe content x, the critical temperature of $Pb_{1-x}Ge_xTe$ increases. Results for $T_c$ obtained by various methods including FIR spectroscopy (Jantsch et al., 1978b), C-V measurements (Jantsch et al., 1981a), anomalies of the high-frequency dielectric constant and the optical energy gap (Jantsch, 1980), and crystal structure analysis (Hohnke et al., 1972) are given in Fig.5.2. For pure PbTe, the transverse optical mode does not become soft (Fig.5.3). Extrapolating $\omega_{TO}^2$ to zero according to (5.1), negative values of $T_c$ are found. In this case, the critical temperature is just a parameter, which characterizes the stability of the cubic phase. For $x > 0.003$ (Fig.5.2), the extrapolated Curie temperature becomes positive, but the actual phase transition does not occur until x exceeds 0.005. In the concentration range $0.003 \leq x \leq 0.005$, $Pb_{1-x}Ge_xTe$ represents an incipient or quantum ferroelectric. The ferroelectric phase transition of incipient ferroelectrics is suppressed by the influence of quantum statistical ("zero-point") fluctuations (Schneider et al., 1976; Müller and Burkhard, 1979; Höchli and Boatner, 1979; Rytz et al., 1980).

Results for the soft mode in the rhombohedral phase of $Pb_{1-x}Ge_xTe$ and SnTe obtained from Raman experiments are given in Figs.2.3-5. As in the incipient case, the $A_1$- and E modes saturate at low temperatures due to zero-point fluctuations. This effect is also observed on paraelectric PbTe (Fig.5.3). The static dielectric constant, which is related to the phonon frequencies by the Lyddane-Sachs-Teller relation (1.1), exhibits the same type of behavior at low temperatures (Figs.3.1,2).

To explain the temperature dependence of the soft mode and the composition dependence of the critical temperature, the soft-mode frequency is considered as a

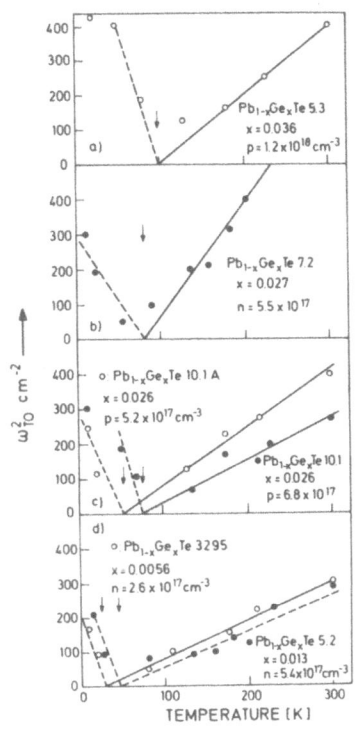

◄ **Fig.5.1.** Squared soft-mode frequency of $Pb_{1-x}Ge_xTe$ versus temperature obtained from FIR reflectivity, after Jantsch et al. (1979)

**Fig.5.2.** Critical temperature of $Pb_{1-x}Ge_xTe$ as a function of alloy composition, obtained from various methods

**Fig.5.3.** Temperature dependence of the soft-mode frequency of PbTe obtained from the FIR method for two samples of different carrier concentrations, after Bauer et al. (1978)

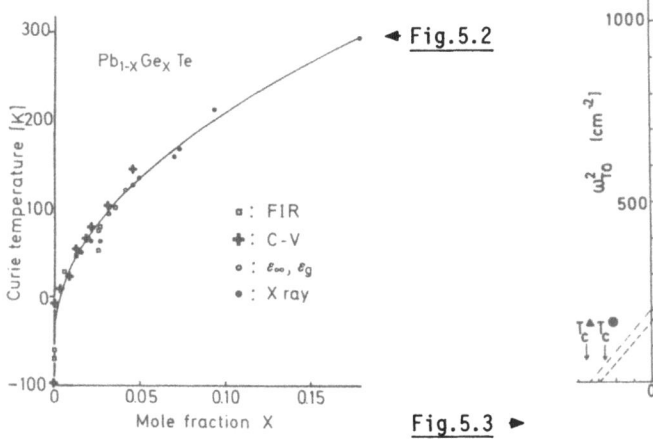

◄ **Fig.5.2**

**Fig.5.3** ►

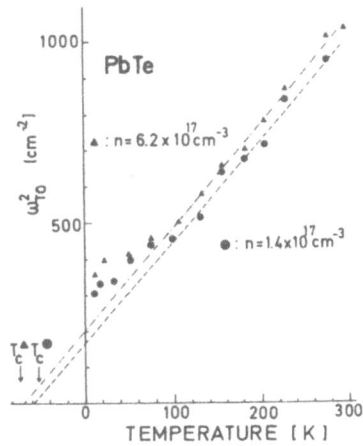

sum of a temperature-independent part describing the composition dependence $\omega_{TO,h}^2(x)$, and a temperature-dependent term $\omega_{ah}^2(T)$, which depends only weakly on composition:

$$\omega_{TO}^2(T,x) = \omega_{TO,h}^2(x) + \omega_{ah}^2(T) \quad . \tag{5.2}$$

Except for the quantum regime at low temperatures, where $\omega_{TO}^2$ becomes temperature independent, $\omega_{ah}^2(T)$ is a nearly linear function of temperature. In the linear region, $\omega_{ah}^2(T)$ is related to the slope of $\omega_{TO}^2$ versus T above and below $T_c$. Experimental

33

Table 5.1. Soft-mode temperature coefficients in the cubic phase ($F_{1u}$ mode) and the rhombohedral phase (E- and $A_1$ modes)

| Material | Compos. x | Carrier conc. (<0 for n-type) $[10^{17}\ cm^{-3}]$ | $T_c$ [K] | $\partial^2\omega_{TO}/\partial T$ $[cm^{-2}/K]$ | | | Reference |
|---|---|---|---|---|---|---|---|
| | | | | $F_{1u}$ | E | $A_1$ | |
| $Pb_{1-x}Ge_xTe$ | 0 | 6.2 | -70 | 2.8 | - | - | Bauer et al. |
| | | 1.4 | -60 | 2.8 | - | - | (1978) |
| | 0.0056 | -2.6 | 29 | 1 | -6.9 | - | Jantsch et al. |
| | 0.013 | -5.4 | 47 | 1.2 | -6.5 | - | (1978b) |
| | 0.026 | 6.8 | 76 | 1.2 | -8.45 | - | |
| | 0.026 | 5.2 | 53 | 1.6 | -5.7 | - | |
| | 0.027 | 5.5 | 81.5 | 3.4 | -3.7 | - | |
| | 0.036 | 12.0 | 100 | 2 | -7.9 | - | |
| | 0.05 | - | 130 | - | -5.2 | -14.6 | Murase and Sugai (1979), Sugai et al. (1979) |
| $Pb_{1-x}Sn_xTe$ | 1 | 1400 | 100 | - | -5.9 | -23.15 | Murase and Sugai (1979) |
| | 1 | 1400-4200 | 100-64 | - | -6.0 | - | Sugai et al. (1979) |
| | 0.37 | | | 3 | | | Dolling and Buyers (1972) |
| | 0.13 | | -110 | 2.5 | | | Daughton et al. |
| | 0.20 | | | 2.4 | | | (1978) |

values for $\partial^2\omega_{TO}/\partial T = \partial^2\omega_{ah}/\partial T$ are given in Table 5.1. The data for components of different composition differ only by a factor of three, although the critical temperature varies between -70 and +140 K, thus justifying an Ansatz like (5.2).

The critical temperature is characterized by

$$\omega_{TO}^2(T_c,x) = 0 \quad . \tag{5.3}$$

Considering the high-temperature cubic phase, which is stabilized by the positive, temperature-dependent part $\omega_{ah}^2(T)$, (5.3) implies a necessary condition for the occurrence of a ferroelectric phase transition:

$$\omega_{TO,h}^2(x) < 0 \quad . \tag{5.4}$$

For $T < T_c$, $\omega_{TO}$ becomes imaginary and the crystal is unstable with respect to this mode. Due to the change in structure it stabilizes. The Landau theory for second-order phase transitions (Landau, 1937; Lines and Glass, 1977) based on a phenomenological expansion of the free energy up to fourth order in a local-order parameter, describes the temperature dependence of the soft mode by

$$\omega_{TO}^2 = B(T - T_c) \qquad \text{for} \quad T > T_c \tag{5.5}$$

$$\omega_{TO}^2 = \zeta B(T - T_c) \qquad \text{for} \quad T < T_c \quad , \tag{5.6}$$

where B is a constant characteristic of the material and $\zeta = 2$. In the low-temperature phase, the slope of $\omega_{TO}^2$ versus T curves should thus be twice the value of the
high-temperature phase and negative. A direct application of this result to the
present case is somewhat complicated by the mode splitting in the rhombohedral
phase, where $\omega_{TO}^2$ in (5.6) should be replaced by an average of the $A_1$- and E-mode
frequencies. Reference to Table 5.1 shows that $\zeta$, the slope ratio of $\omega_{TO}^2$ versus T
of the $F_{1u}$ and the E modes, scatters significantly and that its absolute value
tends to be larger than 2, which may be accounted for by including higher-order
terms within the Landau expansion.

An interband electron-TO-phonon coupling model has been proposed by Kristoffel
and Konsin (1968) in order to explain the temperature dependence of $\omega_{TO}^2$ and the apparent influence of free carriers (Sect.5.3), whereby the former arises mainly from
the temperature dependence of an effective energy gap. However, since the interband
excitation energy is much larger than the thermal energy, the effect on the soft-
mode frequency is too small to account for the observed temperature dependence
(Kawamura, 1980). The temperature dependence of $\omega_{TO}^2$ is therefore attributed to the
anharmonicity of the crystal (Cowley, 1980). In particular, Kawamura (1980) considered fourth-order coupling of the soft-mode to acoustic modes, which increases
the soft-mode frequency. The coupling becomes more effective with increasing temperature due to the increasing phonon population. The experimental data for $\omega_{TO}^2(T)$
was explained by fitting an effective fourth-order phonon-phonon coupling constant
within a Debye model. The same type of approach was originally applied to the zone-
center soft mode of $SrTiO_3$ (Bruce and Cowley, 1973; Cowley, 1980). However, the observed temperature dependence required unrealistically high values of the effective
coupling parameter. A model including only fourth-order phonon-phonon interaction
is thus unsatisfactory. Additional properties, like the nonlinear dielectric constant of $SrTiO_3$, could not be fully explained. Migoni et al. (1976) showed that
second-order Raman scattering and the temperature dependence of the ferroelectric
mode of oxidic perovskites can be explained within a nonlinear shell model by considering a strongly anisotropic and anharmonic intraionic polarizability of the
chalcogen ion. A simplified diatomic linear chain version of this model, which has
been shown to reproduce the essential features of the full 3-dimensional model
(Bilz et al., 1980), has been also applied to the IV-VI compounds PbS, PbSe, PbTe
and SnTe (Bussmann-Holder et al., 1980). This model, which contains 5 phenomenological force-constant parameters, can account for the dispersion of the TA and TO modes
in the rocksalt structure. The temperature dependence of the soft mode was calculated within the self-consistent phonon approximation. In the quantum limit ($T_c \approx 0$)
of the low-temperature regime, a dependence following $\omega_{TO}^2 \sim T^2$ is obtained in agreement with predictions of renormalization group arguments (Schneider et al., 1976;
Oppermann and Thomas, 1975).

In the classical regime, for not too high temperatures, the classical mean-field
behavior, $\gamma = 1$, is recovered in agreement with the Curie-Weiss law. At temperatures

between the classical regime and the low-temperature case (Müller and Burkard, 1979; Rytz et al., 1980), a value of $1.4 < \gamma < 1.8$ is predicted from the self-consistent phonon approximation (Bussmann-Holder et al., 1980). Experimentally, this crossover behavior has been observed on $KTa_{0.9}Nb_{0.1}O_3$ (Kind and Müller, 1976), $SrTiO_3$ (Müller and Burkhard, 1979), and $K_{1-x}Na_xTaO_3$ (Höchli and Boatner, 1979).

For IV-VI compounds, the FIR method is not sufficiently accurate to resolve these rather small deviations from linearity. However, indications of a nonlinear temperature dependence have been found from differential capacitance measurements of Schottky barriers on $Pb_{1-x}Ge_xTe$ (Jantsch et al., 1981a). For $x = 0.003$, which is close to the case of a quantum ferroelectric ($T_c \approx 0$), $\varepsilon_s^{-1}$ shows nearly parabolic behavior. For $x > 0.005$, the phase transition actually occurs. Above $T_c$, the inverse static dielectric constant is not linear in temperature. For $T > T_c + 60$ K the curves bend, attributable to a change of the critical exponent $\gamma$ from its mean field value of 1 at high temperatures to a value of $1 < \gamma < 2$ in the crossover regime. However, a quantitative evaluation of critical exponents from Fig.3.2 is not possible due to additional effects in the static dielectric constant which cause deviations from the LST relation (Sect.5.3).

From the foregoing we conclude that the soft-mode concept applies to the IV-VI compounds at least in the noncritical regime ($\omega_{TO} > 10$ cm$^{-1}$) within experimental accuracy. There are indications, however, of deviations from the soft-mode behavior in the critical regime, which prevent the transverse optical mode frequencies from vanishing at $T_c$ (Sect.5.3). Details of the temperature dependence of the soft mode outside the critical regime can be explained by the nonlinear, anisotropic shell model of Bussmann-Holder et al. (1980).

## 5.2 *Microscopic Origin of Structural Instability and Chemical Trends of Critical Temperature*

The microscopic origin of the crystal structure of IV-VI compounds and their tendency for phase transitions has been attributed to the "resonant" or "mesomeric" nature of chemical bonding (Krebs, 1964; Lucovsky and White, 1973; Littlewood and Heine, 1979). Taking PbTe as a prototype for this class of compounds, bonding occurs through the two 6p electrons of Pb and the four 5p electrons of Te, whereas the fully occupied s states barely contribute (Cohen et al., 1964; Lucovsky and White, 1973; Martinez et al., 1975). Thus, there are six p electrons per atom pair. The sixfold-coordinated cubic NaCl structure is supported by unsaturated p bonds oriented along the cubic [100], [010], and [001] axes. Pair bonds can be formed for only three of the six nearest neighbors. In a linear chain model, two equivalent situations exist according to which neighbor is chosen for these pair bonds.

Indicating pair bonds by a dash (-), these two mesomeric situations are:

Pb-Te  Pb-Te  Pb-Te  ⇌ -Pb  Te-Pb  Te-Pb  Te-  .

In equilibrium, the two situations are indistinguishable and thus resonating (Krebs, 1968; Lucovsky and White, 1973). In response to an optical-type deformation, the bonding electrons easily redistribute to form double bonds. This asymmetry can be stable at low temperatures, thus causing a rhombohedral or orthorhombic distortion of the crystal (Littlewood and Heine, 1979; Kawamura, 1978). The latter structure is favored by s-p hybridization, which occurs in IV-VI compounds of lower molecular weight (GeSe, GeS, SnS, SnSe) (Littlewood, 1980).

An intuitive explanation for the occurrence of soft modes in IV-VI compounds, based on a modified Szigeti model, has been proposed by Grosse (1978) (see also Burstein et al., 1971; Lucovsky et al., 1971). In this model, the soft-mode frequency is decomposed into a spring-constant frequency related to the short-range forces, and a contribution, characterizing the dipole-dipole interaction. Considering empirical values for effective charges, it was shown that the two contributions nearly cancel, in agreement with the soft-mode concept (Cochran, 1960). However, the chemical trends in the series PbTe-SnTe-GeTe cannot be explained by this model. In addition, quantitative models for semiconductors show that $\omega_{TO}$ depends solely on microscopic, short-range potentials and polarizabilities (Porod et al., 1980; Baldereschi, 1980).

Various phenomenological model calculations of the effect of the electron-phonon interaction on the lattice stability of IV-VI compounds have been performed to account for the resonant binding situation (Kristoffel and Konsin, 1968; Kawamura, 1978,1980; Volkov et al., 1976; Natori, 1976). In these models interband electron-phonon coupling (Shukla and Sinha, 1966) in narrow-gap semiconductors is considered responsible for $\omega_{TO,h}^2 < 0$ [see (5.2)].

Introducing an average gap by application of Penn's model (1962), the tendency for structural phase transitions has been attributed to large values of $\varepsilon_\infty$, a large transverse effective charge and a large, phenomenologically introduced interband deformation potential (Katayama and Kawamura, 1977; Kawamura, 1980). However, as pointed out by Littlewood (1979), Penn's model may not be appropriate for the IV-VI compounds due to the large anisotropy of their band structure. Although interband electron-phonon coupling models have successfully explained several other effects like an apparent dependence of $T_c$ on the carrier concentration and on magnetic fields, the observed increasing critical temperature in the series PbTe, SnTe and GeTe has not been satisfactory explained.

A semiempirical theory of the TO frequency, based on the observation that the heat of formation of the cubic IV-VI compounds is related to a pseudopotential form factor, has been proposed by Littlewood (1980). Within this model, negative values for $\omega_{TO,h}^2$ are found for SnTe and GeTe, indicating correctly the instability of the cubic structure for these compounds. The cubic/orthorhombic and cubic/rhombohedral phase boundaries have been discussed in terms of ionicity and covalent parameters, which were derived from a Pauli-force model potential (Littlewood, 1980).

Recently a quantitative pseudopotential model for zone-center phonons was developed by Porod et al. (1980) (PVB model), which does not contain any phenomenological parameters and predicts the observed chemical trends properly. In this model, the harmonic force constants, which determine $\omega_h^2$, are obtained from two contributions: (i) the pure Coulombic interaction of the bare ion cores and (ii) a force mediated by the valence electrons (ten electrons per ion pair in the present case), which are redistributed by the phononic deformation of the lattice. The latter is approximated by a static lattice distortion. The core-electron interaction outside the core is Coulombic, but inside the core it is modified as described by tabulated electron-ion pseudopotentials. The phonon-induced redistribution of electrons is described by the microscopic dielectric function $\varepsilon(\underline{q} + \underline{G}, \underline{q} + \underline{G}')$ in Fourier space. This model has been applied to various III-V, II-VI and IV-VI semiconductors. The results for $\omega_{TO}$ have been shown to agree with experimental data within 15%.

In the case of IV-VI compounds, the two terms (i) and (ii) nearly cancel, thus explaining the instability of the lattice. The chemical trends are mainly caused by the reduced mass of the ion pair and the atomic potential, reflecting the fact that lighter elements attract the outer electrons more strongly. The experimentally observed trend of an increasing critical temperature in the series PbTe, SnTe and GeTe is fully explained by this theory. Thus, IV-VI compounds represent a rare example of ferroelectrics giving a basic understanding of the mechanisms responsible for lattice instability. The PVB model is so applicable to IV-VI compounds because of the simplicity of their crystal structure and the well-known pseudo-potential form factors, which have not been determined accurately for ionic solids. The energy gap does not enter in the PVB model and has been shown to be unimportant for the lattice instability, in contrast to the phenomenological treatment of interband electron-phonon coupling models.

Alloying effects deserve further attention. The critical temperature of $Pb_{1-x}Ge_xTe$ and $Pb_{1-x}Sn_xTe$ exhibits a nonlinear variation with the composition parameter x: on the PbTe-rich side, the variation of $T_c$ is much stronger than for large values of x. Especially in $Pb_{1-x}Ge_xTe$, the Curie temperature increases drastically from about ~70 K to 220 K between x = 0 and x = 0.1 (Fig.5.2). To explain this behavior, Katayama and Murase (1980) proposed a model based on the assumption of a local instability of dilute Ge ions on regular Pb sites. This instability may be associated with the relatively large differences in the ionic radii of $Pb^{++}$ and $Ge^{++}$ (Katayama and Murase, 1980; Yaraneri et al., 1982). The off-center Ge ions produce local dipole moments whose interaction may lead to a cooperative Jahn-Teller - like effect, so increasing the tendency for a bulk distortion (Katayama and Murase, 1980).

## 5.3 Influence of Lattice Defects

An influence of lattice point defects on the crystal structure and its transition temperature and hence on the dielectric properties has been observed in GeTe (Schubert and Fricke, 1951), SnTe (Muldawer, 1973; Iizumi et al., 1975; Valassiades and Economou, 1975; Kobayashi et al., 1976; Sugai et al., 1977b), $Pb_{1-x}Sn_xTe$ (Takasaki and Tanaka, 1977) and $Pb_{1-x}Ge_xTe$ (Hohnke et al., 1972) and PbTe (Jantsch and Lopez-Otero, 1976; Bauer et al., 1978). Lattice defects may arise from non-stoichiometry of the binary- or pseudo-binary compounds (Brebrick, 1962) or as native defects during crystal growth or subsequent thermal treatment (Lopez-Otero, 1978). The determination of defect concentrations in IV-VI compounds presents a difficult problem (Sect.3.1). However, since the concentration of free carriers depends on the defect concentration (3.10), their influence can be studied at least qualitatively by investigating samples with different carrier concentration.

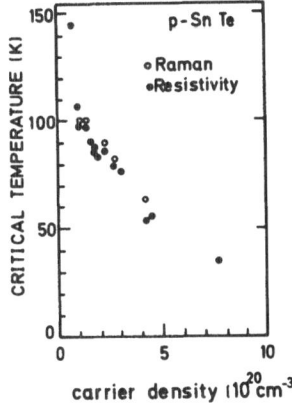

**Fig.5.4.** Critical temperature of SnTe versus hole concentration obtained from Raman experiments (Sugai et al., 1977) and the resistivity anomaly (Kobayashi et al., 1976; Kawamura, 1978; Grassie et al., 1979)

For SnTe only specimens with low carrier concentration ($p \sim 10^{20}$ $cm^{-3}$) were found to exhibit a structural phase transition. Results for the critical temperature of SnTe, obtained by various methods, are given in Fig.5.4. With increasing carrier concentration, the critical temperature decreases, the cubic structure is stabilized: PbTe remains cubic even for the lowest available carrier concentrations of about $10^{16}$ $cm^{-3}$. The same tendency is evident from the increase of the soft-mode frequency. In Fig.5.3, FIR results for $\omega_{TO}^2$ as a function of temperature are given for two samples of different carrier concentration. The curve for the sample with higher carrier concentration appears to be shifted towards higher values of $\omega_{TO}^2$. The extrapolated Curie temperature decreases with increasing defect concentration. In principle, the static dielectric constant shows the same type of behavior (Fig.3.1) but more pronounced due to an additional shift of the inverse quasi-static dielectric constant, $\varepsilon_s^{-1}$, relative to the Lyddane-Sachs-Teller value $\varepsilon_0^{-1}$. This effect is evident from Fig.3.3 (see also discussion in Sect.3.1).

The influence of lattice defects on the soft mode and the critical temperature has been attributed (Kobayashi et al., 1976) to the band population by defect-induced free carriers in the interband electron-phonon coupling model (Kawamura et al., 1974; Kobayashi et al., 1976; Kawamura, 1980). To explain the influence of free carriers, an electron-TO-phonon interaction mediated by virtual excitations of electrons from the vicinity of the valence band edge to the conduction band has been suggested. With increasing energy gap the strength of the interaction decreases. In the presence of high carrier concentrations, the effective energy gap increases due to the degenerate electron or hole concentration (Kawamura, 1980) analogous to the Burstein-Moss shift of the fundamental adsorption. The interband mechanism causes a negative renormalization of the TO-mode frequency in the cubic phase, thus explaining the lattice instability condition $\omega^2_{TO,h} < 0$, (5.4). This negative contribution becomes weaker with increasing carrier concentration, resulting in an increase of $\omega^2_{TO}$ in the cubic phase and a decrease of $T_c$. The coupling strength is expressed in terms of a phenomenologically introduced interband optical deformation potential constant, which is fitted to the observed dependence of $T_c$ on the carrier concentration. Literature values for this deformation potential scatter considerably. For SnTe values ranging from 2.3 eV (Kawamura et al., 1975) to 25 eV (Kawamura, 1978) have been found, indicating that other mechanisms are important when considering the influence of defects.

In addition, the exact, microscopic treatment of electron-phonon interaction in the PVB model shows that the energy gap plays no role in the lattice dynamics of IV-VI compounds. The influence of free carriers due to band population turns out to be negligible within this theory (Vogl and Vergês, 1982). Experimental support for this conclusion was derived from investigations of the influence of hydrostatic pressure on $T_c$ (Suski et al., 1981).

The stabilization of the soft mode by lattice point defects is not restricted to narrow-gap IV-VI semiconductors. Both the zone-boundary soft-mode frequency of SrTiO$_3$ ($T_c$ = 105 K for crystals with low defect concentrations) and the zone-center soft-mode frequency of SrTiO$_3$ ($T_c < 0$) were observed to increase upon introduction of oxygen vacancies (Wagner et al., 1980; Bäuerle et al., 1980). The antiferrodistortive phase transition at 105 K is completely suppressed for oxygen-vacancy concentrations higher than 2%. The situation is quite similar to that encountered for IV-VI compounds, chalcogen vacancies acting as donors (Perluzzo and Destry, 1978). Again the high-temperature phase is stabilized either by free carriers or by vacancies, which act as a source of free carriers. The observed vacancy-induced shifts in the zone-center soft mode frequencies of SrTiO$_3$ have been interpreted in terms of the nonlinear shell model. The experimental data can be explained by considering the effect of screening by free electrons on the attractive Ti-O interaction (Bussmann-Holder et al., 1981). The decrease of the dipolar Coulomb part of the harmonic force constants leads to an increase of the soft-mode frequency of the same order of magnitude as the experimental value of $\Delta\omega^2 \approx 0.58 \cdot 10^4$ cm$^{-2}$/atom% oxygen vacancies.

40

In the case of IV-VI compounds, additional effects besides the mode shift occur, which are obviously related to defects: experimental literature values for the static dielectric constant $\varepsilon_s$ scatter by nearly two orders of magnitude (Foley and Langenberg, 1977) and disagree systematically with $\varepsilon_0$, the value obtained from the optical mode frequencies via the Lyddane-Sachs-Teller relation (1.1) (see also Sect.3.1 and Fig.3.3).

An apparent violation of the Lyddane-Sachs-Teller relation, which contains only contributions from the optical phonon oscillator strengths (1.1), has been discussed also for other displacive ferroelectrics like $BaTiO_3$, $LiTiO_3$, $LiTaO_3$ and $Pb_{1-x}La_xTi_{1-x/4}O_3$ (Burns, 1976). For these compounds the subject is still controversial because of difficulties in evaluating soft-mode frequencies from Raman- or FIR experiments (Barker and Hopfield, 1964; Barker, 1967; Servóin et al., 1980). Problems arise from the frequency dependence of the phonon damping constant, $\Gamma_j(\omega)$, of the classical oscillator dielectric function, given in its simplest one-mode form in (2.4). In this situation, the maximum of the imaginary part of the dielectric function $\varepsilon''(\omega)$, does not coincide in frequency with the pole of its real part $\varepsilon'(\omega)$ (Barker, 1967). The LST relation is still valid if $\omega_{TO}^2$ in (1.1) is considered as an effective low-frequency restoring force constant $\omega_F^2$ (Barker, 1967), which is obtained by fitting a classical one-mode oscillator to the low-frequency end of the FIR reflectivity spectrum (Sect.2.4). In contrast, assigning $\omega_{TO}$ to peaks in $\varepsilon''(\omega)$, which dominate transmission or Raman spectra, may lead to incorrect results for the mode frequencies.

In the present case the soft-mode frequency (Fig.2.2) was determined from FIR reflectivity close to the TO resonance. Hence, we conclude that the observed deviations from the LST relation (Fig.2.2) are due to an additional effect, which causes dispersion in the frequency range between the Reststrahlen regime and the low-frequency MHz regime, where the quasi-static dielectric constant is measured.

In Burn's (1976) model, additional dispersion is caused by a phenomenologically introduced low-frequency mode. The oscillator strength of this "defect mode" is assumed proportional to the concentration of defects. The two oscillators (the defect oscillator and the TO mode) are coupled through the microscopic polarization, in analogy to the Szigeti (1949) model. For the dielectric function, $\tilde{\varepsilon}(\omega)$, the following expression is obtained (Burns, 1976; Bauer et al., 1978):

$$\tilde{\varepsilon}(\omega) = \tilde{\varepsilon}_{ph} + \tilde{\chi}_{FC} + \tilde{\chi}_D = \tilde{\varepsilon}_{ph} + \tilde{\chi}_{FC} + \frac{\Omega_D(\tilde{\varepsilon}_{ph}+2)^2}{\omega_D^2 - \omega^2 + i\omega\Gamma_D - \Omega_D^2(\tilde{\varepsilon}_{ph}+2)} , \tag{5.7}$$

with the defect oscillator strength given by:

$$\Omega_D^2 = \frac{4\pi N_D Q_D^2}{9M_D} . \tag{5.8}$$

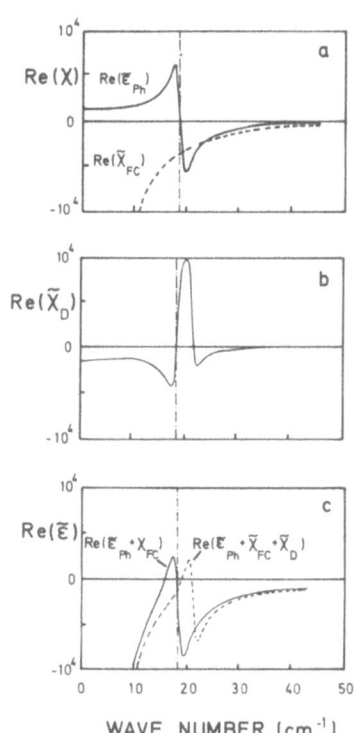

WAVE NUMBER (cm⁻¹)

**Fig.5.5.** a) Real part of the phonon contribution to the dielectric function ($\tilde{\varepsilon}_{ph}$) and of the free-carrier susceptibility $\tilde{\chi}_{FC}$, according to (2.2). The parameters used (given in Table 5.2) correspond to values typical for PbTe. b) Defect susceptibility (real part) as a function of frequency according to the third term in (5.7) (parameter values in Table 5.2). c) Total dielectric function (real part) without (solid line) and with (dashed line) defect contribution

**Table 5.2.** Parameter values used for Fig.5.5

| Phonon | Free carriers | Defect oscillator |
|---|---|---|
| $\omega_{TO} = 18.5$ cm$^{-1}$ | $N = 5.2 \cdot 10^{17}$ cm$^{-3}$ | $\omega_D = 4$ cm$^{-1}$ |
| $\Gamma = 2$ cm$^{-1}$ | $m_p^* = 0.04$ | $\Gamma_D = 0.5$ cm$^{-1}$ |
| $\varepsilon_\infty = 34.5$ | $\omega_\tau = 0.5$ cm$^{-1}$ | $M_D = 200$ a.m.u. |
| $\omega_{LO} = 114$ cm$^{-1}$ | | $N_D = 1 \cdot 10^{18}$ cm$^{-3}$ |

Here $N_D$, $Q_D$ and $M_D$ stand for the concentration, the effective charge and mass of the defects, respectively, $\omega_D$ denotes the resonance frequency and $\Gamma_D$ the damping constant of the defect oscillator. The free-carrier susceptibility $\tilde{\chi}_{FC}$ and the pure phonon dielectric function $\tilde{\varepsilon}_{ph}$ were given in (2.2,5), respectively. In Fig. 5.5 the real part of $\tilde{\varepsilon}_{ph}$ and the defect susceptibility $\tilde{\chi}_D$ are given together with the free-carrier contribution $\tilde{\chi}_{FC}$, according to (5.7). The parameters used are given in Table 5.2.

The introduction of $\tilde{\chi}_D$ has two main effects: (i) the static dielectric constant $\varepsilon_s = \varepsilon_0 + \text{Re}\{\chi_D(\omega \to 0)\}$ is modified and (ii) the resonance of $\text{Re}\{\tilde{\varepsilon}(\omega)\}$ close to the soft-mode frequency $\omega_{TO}$ of the perfect crystal is shifted towards a higher value $\omega_{TO}^*$. For the parameters used in Fig.5.5 (Table 5.2), the contribution of the defect oscillator (i) for the low-frequency dielectric constant $\text{Re}\{\chi_D(\omega \to 0)\}$ is negative down to low frequencies thus causing a positive shift of $\varepsilon_s^{-1}$ relative to the LST result $\varepsilon_0^{-1}$, as observed experimentally (Fig.3.3). The stabilization of the soft mode (ii) is evident from Fig.5.4: the pole at $\omega_{TO}$, which is smeared out due to damping, occurs at a higher frequency. The shift of $\omega_{TO}$ increases with $\Omega_D$ and hence with the concentration of defects. The temperature dependence of $\omega_{TO}^*$ is modified as compared to the perfect crystal: for temperatures far away from $T_c$, where $\omega_{TO}^2 \gg \omega_D^2$, the concentration-dependent interaction of the soft mode with the defect mode causes only a small renormalization of $\omega_{TO}$, whereas for $\omega_{TO} \approx \omega_D$ the two modes "repel" each other strongest, as is usual in mode-coupling problems. This effect may explain the scattering of literature values of the temperature coefficients $\partial\omega_{TO}^2/\partial T$ (Table 5.1) as a consequence of defects differing in concentration from sample to sample.

Thus, in principle, Burn's phenomenological model may account for the observed anomalies of the static dielectric constant and defect-induced changes of the soft-mode frequency. Predictions for the critical regime, particularly for the concentration dependence of $T_c$, cannot be made, since no criteria for the stability of a phase are contained within this model. Experimental data for $\varepsilon_0^{-1}$, as given in Fig. 3.3, can be fitted using Burn's model (Mitter, 1981; Jantsch, 1982). In the absence of a microscopic model for the defect mechanism, however, only a qualitative discussion is possible. The microscopic origin of the low-frequency oscillator remains to be identified. For antimony sulfoiodide, which exhibits a ferroelectric phase transition close to room temperature, anomalous dispersion has been found in the frequency range of 0.1-1 cm$^{-1}$ (Grigas and Karpus, 1967; Grigas and Beliackas, 1978; Irie, 1978). On the other hand, a strong quasi-elastic component in the Raman spectra of SbSI close to $T_c$ has been interpreted as a central peak which couples with the soft mode (Steigmeier et al., 1971,1975). A central peak in SbSI may have also been observed in inelastic neutron scattering experiments (Pouget et al., 1979). The results from a coupled soft- and central-mode oscillator model fitted to the Raman spectra are consistent with the anomalous dispersion in the μ-wave regime (Steigmeier et al., 1975). Thus, it is tempting to assign the anomalous low-frequency dielectric properties of SbSI to the influence of a central peak.

Critical phenomena and in particular the occurrence of a central peak have received considerable attention recently (Lines and Glass, 1977; Bruce and Cowley, 1980). One class of theories explains the central peak phenomenon in terms of coupling of the soft mode with a slowly relaxing mode (Schwabl, 1972) which originates from defects (Axe and Shirane, 1974; Halperin and Varma, 1976; Schmidt and Schwabl, 1977,1978; Bruce and Cowley, 1980). The order-parameter susceptibility obtained is similar in structure to the susceptibility in Burns model (5.7). Thus the essential results are similar to those described above. In addition, predictions have been made for the defect-induced shift of $T_c$. For defects which couple quadratically with the order parameter, a linear shift of $T_c$ with the concentration $N_D$ is obtained from lowest-order perturbation theory (Schmidt and Schwabl, 1977, 1978) in agreement with experimental results for the zone boundary mode of heavily reduced $SrTiO_3$ (Wagner et al., 1980).

A preliminary report on the observation of a central peak in SnTe by Raman scattering is rather unconvincing (Sugai et al., 1977a). Further investigations of the central peak and the influence of defects, whose origin and structure remain to be characterized, are required for a satisfactory understanding of this complex subject.

## 5.4 Influence of Magnetic Fields

An influence of high magnetic fields on the critical temperature of $Pb_{1-x}Ge_xTe$ (Murase et al., 1976) and $Pb_{1-x}Sn_xTe$ (Takaoka and Murase, 1979) has been found rather indirectly from a field-induced temperature shift of the resistance anomaly peak (Sect.4.2). For $Pb_{1-x}Ge_xTe$ (x = 0.01), $T_c$ was observed to increase by 1 to 3 K for field strengths of 8 T. In Fig.5.6 results for the temperature dependence of the inverse static dielectric constant $\varepsilon_s^{-1}$, obtained from C-V measurements on $Pb_{1-x}Ge_xTe$ (x = 0.02), are given with and without a magnetic field. In the presence of a magnetic field of 8.2 T an increase of the extrapolated critical temperature by 1.2 K is observed (Jantsch et al., 1980), which is comparable to the results obtained from the resistance anomaly. Experimental results for $Pb_{1-x}Sn_xTe$ close to the zero-gap situation (x = 0.4) are conflicting. From resistance anomaly measurements a change of $T_c$ by up to 10 K at 8 T was found, which has been interpreted in terms of the interband electron-phonon coupling model (Takaoka and Murase, 1979; Litvinov et al., 1979; Volkov and Litvinov, 1980). On the other hand, in investigations of the static dielectric constant by means of magnetoplasma wave experiments (Sect.3.2) up to field strengths of 25 T, no evidence of magnetic fields on $T_c$ could be found (Nishi et al., 1980). These latter results are inconsistent with the assumption of an appreciable contribution of interband electron-phonon interaction due to electronic states in the neighborhood of the minimum energy gap (Takaoka and Murase, 1979).

An influence of high magnetic fields on the ferroelectric phase transition has been also found in $BaTiO_3$ from investigations of the static dielectric constant (Wagner and Bäuerle, 1981). In the presence of a magnetic field of 20 T the minimum of the inverse static dielectric constant at the critical temperature is shifted upwards in temperature by 0.2 K to 0.3 K, depending on the relative orientation of the magnetic field and the electric field employed in the capacitance measurements. A tentative explanation of this effect was given in terms of a magnetic-field-induced squeezing of the oxygen 2p- and titanium 3d orbitals (Wagner and Bäuerle, 1981). So far no rigorous microscopic treatment of these interesting and completely new effects has been given in the literature.

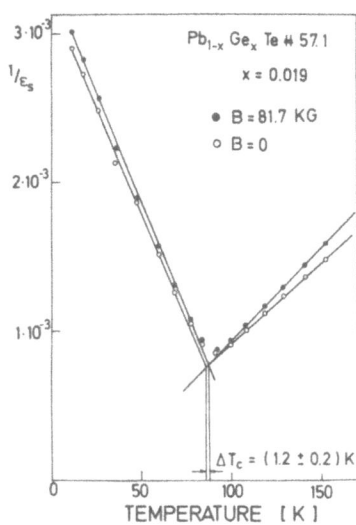

**Fig.5.6.** Inverse static dielectric constant of $Pb_{1-x}Ge_xTe$ obtained from differential capacitance measurements as a function of temperature with (full dots) and without (circles) a magnetic field applied perpendicular to the (111) surface plane, after Jantsch et al. (1980)

## 6. Summary

The narrow-gap semiconductors PbTe, SnTe, GeTe and their alloys exhibit a tendency for a ferroelectric displacive phase transition from a high-temperature rocksalt structure to a rhombohedral phase at low temperatures. The critical temperatures vary within an exceedingly wide range: -70 K(PbTe) up to 650 K(GeTe). Owing to the outstanding simplicity of their crystal structure and their well-known electronic properties, the instability and the chemical trends of $T_c$ can be understood in terms of a quantitative pseudopotential model (Sect.5.2). Phenomenological models explain the temperature dependence of the soft mode outside the critical regime fairly well (Sect.5.1). Critical phenomena, however, deserve further attention: the anomalies of the static dielectric constant (Sects.3.1 and 5.3) and the influence of defects are possibly related to a central peak. Additional systematic investigations on the nature and influence of defects are required to obtain more insight into this highly interesting field of general importance.

*Acknowledgements.* Much of the work presented in this paper was done in collaboration with G. Bauer and A. Lopez-Otero and their contribution is gratefully acknowledged. I also wish to thank H. Heinrich and H.J. Queisser for their continuous interest and encouragement, as well as A. Martin and H. Vogt for many suggestions and their thorough revision of the manuscript. My thanks are also due to H. Bilz, A. Bussmann-Holder and D. Schmeltzer for many helpful discussions.

# References

Abstreiter, G., Trommer, R., Cardona, M., Pinczuk, A. (1979): Solid State Commun. **30**, 703

Allgaier, R.S., Scanlon, W.W. (1958): Phys. Rev. **111**, 1929

Allgaier, R.S., Houston, B.B. (1962): Proc. VIth Int. Conf. of Semiconductors, Exeter (Institute of Physics and Physical Society, London) p. 172

Alperin, H.A., Pickart, S.J., Rhyne, J.J., Minkiewicz, V.J. (1972):Phys. Lett. **40**A, 295

Anderson, P.W. (1960): In *Fizika Dielectrikov*, ed. by G.I. Skanavi (Moscow, Acad. Nauk, SSR)

Antcliffe, G.A., Bate, R.T., Buss, D.D. (1973): Solid State Commun. **13**, 1003

Axe, J.D., Shirane, G. (1974): Phys. Rev. B**8**, 1965

Baldereschi, A. (1980): Proc. 15th Int. Conf. Physics of Semiconductors, Kyoto, ed. by S. Tanaka, Y. Toyozawa (Physical Society Japan, Tokyo): J. Phys. Soc. Japan **49**, Suppl. A, p. 155

Barker, A.S., Jr., Hopfield, J.J. (1964): Phys. Rev. **135**, A 1732

Barker, A.S., Jr. (1967): In *Ferroelectricity*, ed. by E.F. Weller (Elsevier) p. 231

Barrett, J.H. (1952): Phys. Rev. **86**, 118

Bate, R.T., Carter, D.L., Wrobel, J.S. (1970): Phys. Rev. Lett. **25**, 159; Proc. 10th Int. Conf. Phys. of Semiconductors, ed. by S.P. Keller, J.C. Hensel, F. Stern (United States Atomic Energy Commission, Cambridge, Mass.) p. 125

Bauer, G., Burkhard, H., Jantsch, W., Unterleitner, F., Lopez-Otero, A. (1978): Proc. Int. Conf. on Lattice Dynamics, Paris, ed. by M. Balkanski (Flammarion Sciences) p. 669

Bauer, G. (1978): Proc. Int. Conf. on the Application of High Magnetic Fields in Semiconductor Physics, ed. by J.F. Ryan (Clarendon, Oxford) p. 153

Bauer, G. (1980): In *Narrow Gap Semiconductors, Physics and Applications*, Lecture Notes in Physics, Vol. 133, ed. by W. Zawadzki (Springer, Berlin, Heidelberg, New York) p. 407; and in *Applications of High Magnetic Fields in Semiconductor Physics*, Lecture Notes in Physics, Vol. 177, ed. by G. Landwehr (Springer, Berlin, Heidelberg, New York 1983) p. 259

Bäuerle, D., Wagner, D., Wöhlecke, M., Dorner, B., Kraxenberger, H. (1980): Z. Phys. B**38**, 335

Bierly, J.N., Muldawer, L., Beckmann, O. (1963): Acta Metall. **11**, 447

Bilz, H., Bussmann, A., Benedek, G., Büttner, H., Strauch, D. (1980): Ferro-electrics **25**, 339

Brebrick, R.F. (1962): J. Phys. Chem. Solids **24**, 27

Brillson, L.J., Burstein, E. (1971): Phys. Rev. Lett **27**, 808

Brillson, L.J., Burstein, E., Muldawer, L. (1974): Phys. Rev. B**9**, 1547

Bruce, A.D., Cowley, R.A. (1973): J. Phys. C**6**, 2422

Bruce, A.D., Cowley, R.A. (1980): Adv. Phys. **29**, 219

Burkhard, H., Bauer, G., Lopez-Otero, A. (1976): Solid State Commun. **18**, 773

Burkhard, H., Bauer, G., Lopez-Otero, A. (1977): J. Opt. Soc. Am. **67**, 946

Burkhard, H., Bauer, G., Lopez-Otero, A. (1978): Phys. Rev. B**18**, 2935

Burns, G. (1976): Phys. Rev. B**13**, 215

Burns, G., Burstein, E. (1974): Ferroelectrics **7**, 297

Burstein, E., Pinczuk, A., Wallis, R.F. (1971): In *The Physics of Semimetals and Narrow-Gap Semiconductors*, ed. by D.L. Carter, R.T. Bate (Pergamon) p. 251

Buss, D.D., Kinch, M.A. (1973): J. Nonmetals **1**, 111

Bussmann-Holder, A., Bilz, H., Kress, W. (1980): Proc. 15th Int. Conf. Phys. of Semiconductors, Kyoto, ed. by S. Tanaka, Y. Toyozawa (Physical Society Japan, Tokyo): J. Phys. Soc. Japan **49**, Suppl. A, p. 737

Bussmann-Holder, A., Bilz, H., Bäuerle, D., Wagner, D. (1981): Z. Phys. B**41**, 353

Cape, J.A., Hale, L.G., Tennant, W.E. (1977): Surface Science **62**, 639

Cardona, M. (1968): In *Solid State Physics, Nuclear Physics and Particle Physics*, Ninth Latin American School of Physics (W.A. Benjamin INC., New York) p. 738

Clarke, R. (1978): Phys. Rev. B**18**, 4920

Claus, R., Merten, L., Brandmüller, J. (1975): Springer Tracts in Modern Physics, Vol. 75, ed. by G. Höhler (Springer, Berlin, Heidelberg, New York)

Cochran, W. (1960): Adv. Phys. **9**, 387
Cochran, W., Cowley, R.A. (1962): J. Phys. Chem. Sol. **23**, 447
Cochran, W. (1964): Phys. Lett. **13**, 193
Cochran, W., Cowley, R.A., Dolling, G., Elcombe, M.M. (1966): Proc. Roy. Soc. (London) **A293**, 433
Cohen, M.H., Falicov, L.M., Golin, S. (1964): IBM J. Res. Dev. **8**, 215
Cowley, R.A., Dolling, G. (1965): Phys. Rev. Lett. **14**, 549
Cowley, E.R., Darby, J.K., Pawley, G.S. (1969): J. Phys. C: Solid State Phys. Ser. 2, **2**, 1916
Cowley, R.A. (1980): Adv. Phys. **29**, 1
Dalven, R. (1973): In *Solid State Physics*, Vol. 28, ed. by H. Ehrenreich, F. Seitz, D. Turnbull (Academic, New York, London) p. 171
Damen, T.C., Porto, S.P.S., Tell, B. (1966): Phys. Rev. **142**, 570
Daughton, W.J., Tompson, C.W., Gürmen, E. (1978): J. Phys. C: Solid State Phys. **11**, 1573
Devlin, G.E., Davis, J.L., Chase, L., Geschwind, S. (1971): Appl. Phys. Lett. **19**, 138
Dimmock, J.O., Melngailis, I., Strauss, A.J. (1966): Phys. Rev. Lett. **16**, 1193
Dionne, G., Woolley, J.C. (1972): Phys. Rev. **B6**, 3898
Dolling, G., Buyers, W.J.L. (1973): J. Nonmetals **1**, 159
Finkenrath, H., Köhler, H. (1966): Phys. Lett. **23**, 437
Foley, G.M.T., Langenberg, D.N. (1976): Solid State Commun. **18**, 35
Foley, G.M.T., Langenberg, D.N. (1977): Phys. Rev. **B15**, 4830
Gerlach, E., Grosse, P. (1977): *Festkörperprobleme*, Vol. 17, ed. by J. Treusch (Vieweg, Braunschweig) p. 157
Grassie, A.D.C., Agapito, J.A., Gonzales, P. (1979): J. Phys. C: Solid State Phys. **12**, L925
Gresslehner, K.H., Jantsch, W., Lischka, K., Palmetshofer, L., Heinrich, H. (1978): Proc. III. Int. Conf. on Narrow-Gap Semiconductors, ed. by J. Rauluskiewiez, M. Górska, E. Kaczmarek (PWN-Polish Scientific Publishers, Warsaw) p. 205
Grishechkina, S.P., Zhokhovets, S.V., Kopylovskii, B.D., Shotov, A.P. (1978): Fiz. Tekh. Poluprovodn. **12**, 1132 [English transl.: Sov. Phys. Semicond. **12**, 671]
Grigas, I.P., Karpus, A.S. (1967): Fizika Tverd. Tela **9**, 2887 [English transl.: Sov. Phys.-Sol. State **9**, 2270 (1968)]
Grigas, J., Beliackas, R. (1978): Ferroelectrics **19**, 113
Grosse, P. (1978): Proc. III. Int. Conf. of Narrow-Gap Semiconductors, ed by J. Rauluskiewicz, M. Górska, E. Kaczmarek (PWN-Polish Scientific Publishers, Warsaw) p. 41
Haas, D., Heinrich, H., Jantsch, W., Lischka, K., Lopez-Otero, A., Palmetshofer, L., Wagenhuber, M. (1975): Critical Reviews in Solid State Sciences **5**, 547
Halperin, B.I., Varma, C.M. (1976): Phys. Rev. **B14**, 4030
Harbeke, G., Steigmeier, E.F. (1980): Private communication
Hatta, I., Kobayashi, K.L.I. (1977): Solid State Commun. **22**, 775
Hatta, I., Rehwald, W. (1977): J. Phys. C: Solid State Phys. **10**, 2075
Heinrich, H., Lopez-Otero, A., Palmetshofer, L., Haas, L.D. (1975): Lattice Defects in Semiconductors 1974, Honorary Editor: F.A. Huntley, Inst. Phys. Conf. Ser. Number **23**, p. 264 (Institute of Physics, London, Bristol)
Heinrich, H. (1980): In *Narrow-Gap Semiconductors, Physics and Applications*, Lectures Notes in Physics, Vol. 133, ed. by W. Zawadzki (Springer, Berlin, Heidelberg, New York) p. 407
Höchli, U.T., Boatner, L.A. (1979): Phys. Rev. **B20**, 266
Hohnke, D.K., Holloway, H., Kaiser, S. (1972): J. Phys. Chem. Sol. **33**, 2053
Holloway, H. (1980): Thin Solid Films **11**, 105
Ichiguchi, T., Nishikawa, S., Murase, K. (1980): Solid State Commun. **34**, 309
Iizumi, M., Hamaguchi, Y., Komatsubara, K.F., Kato, Y. (1975): J. Phys. Soc. Japan **38**, 443
Irie, K. (1978): Ferroelectrics **21**, 395
Jantsch, W., Lopez-Otero, A. (1976): Proc. 13th Int. Conf. Phys. of Semiconductors, ed. by F.G. Fumi (Tipografica Marves, Rome) p. 487
Jantsch, W., Rozenbergs, J., Heinrich, H. (1978a): Solid-State Electronics **21**, 103
Jantsch, W., Lopez-Otero, A., Bauer, G. (1978b): Infrared Phys. **18**, 877

Jantsch, W., Bauer, G., Lopez-Otero, A. (1979): Proc. 14th Int. Conf. Phys. of Semiconductors, ed. by B.L.H. Wilson (The Institute of Physics, Bristol and London) p. 445

Jantsch, W. (1980): Z. Phys. B**40**, 193

Jantsch, W., Stolz, H.J. (1980): Unpublished

Jantsch, W., Mitter, H., Lopez-Otero, A. (1980): Unpublished

Jantsch, W., Mitter, H., Lopez-Otero, A. (1981a) Z. Phys. B - Condensed Matter **41**, 287

Jantsch, W., Bauer, G., Krost, A., Lopez-Otero, A. (1981b): Ferroelectrics **38**, 905

Jantsch, W. (1982): In *Physics of Narrow Gap Semiconductors*, Lecture Notes in Physics, Vol. 152, ed. by E. Gornik, H. Heinrich, L. Palmetshofer (Springer, Berlin, Heidelberg, New York) p. 226

Kanai, Y., Shohno, K. (1963): Jap. J. Appl. Phys. **2**, 6

Katayama, S. (1976): Solid State Commun. **19**, 381

Katayama, S., Kawamura, H. (1977): Solid State Commun. **21**, 521

Katayama, S., Mills, D.L. (1979): Phys. Rev. B**19**, 6372

Katayama, S., Mills, D.L. (1980): Phys. Rev. B**22**, 336

Katayama, S., Murase, K. (1980): Solid State Commun. **36**, 707; Murase, K.: Proc. 15th Int. Conf. Phys. of Semiconductors, Kyoto, ed. by S. Tanaka, Y. Toyozawa (Physical Society Japan, Tokyo): J. Phys. Soc. Japan **49**, Suppl. A, p. 725

Kawamura, M., Katayama, S., Takano, S., Hotta, S. (1974): Solid State Commun. **14**, 259

Kawamura, H., Murase, K., Nishikawa, S., Nishi, S., Katayama, S. (1975): Solid State Commun. **17**, 341

Kawamura, H. (1978): Proc. Int. Conf. of Narrow-Gap Semiconductors, ed. by J. Rauluskiewicz, M. Górska, E. Kaczmarek (PWN-Polish Scientific Publishers, Warsaw) p. 7

Kawamura, H., Murase, K., Sugai, S., Takaoka, S., Nishikawa, S., Nishi, S., Katayama, S. (1978): Proc. Int. Conf. on Lattice Dynamics, Paris, ed. by M. Balkanski (Flammarion Sciences) p. 658

Kawamura, H. (1980): In *Narrow-Gap Semiconductors, Physics and Applications*, Lecture Notes in Physics, Vol. 133, ed. by W. Zawadzki (Springer, Berlin, Heidelberg, New York) p. 470

Kind, R., Müller, K.A. (1976): Communications on Physics **1**, 223

Kobayashi, K.L.I., Kato, Y., Katayama, Y., Komatsubara, K.F. (1975): Solid State Commun. **17**, 875

Kobayashi, K.L.I., Kato, Y., Katayama, Y., Komatsubara, K.F. (1976): Phys. Rev. Lett. **37**, 772

Komatsubara, K.F., Kato, Y., Kobayashi, K.L.I., Iizumi, M., Hamaguchi, Y. (1974): Proc. 12th Int. Conf. Phys. of Semiconductors, ed. by M.H. Pilkuhn (Teubner, Stuttgart) p. 602

Krebs, H. (1968): *Grundzüge der anorganischen Kristallchemie* (Ferdinand Enke, Stuttgart) p. 183f

Kristoffel, N., Konsin, P. (1968): Phys. Stat. Sol. **28**, 731; Konsin, P. (1982): Ferroelectrics **45**, 45

Landau, L.D. (1937): Phys. Z. Sowjun. **11**, 26, 545; Zh. eksp. teor. Fiz. **7**, 19, 627

Larkin, A.I., Khmel'nitskii, D.E. (1969): Sov. Phys.-JETP **29**, 1123

Lehmann, H., Nimtz, G., Haas, L.D., Jakobus, T. (1981): Applied Physics **25**, 291

Lewis, A.V., Nicholas, R.J., Ramage, J.C., Bauer, G., Lopez-Otero, A., Stradling, R.A. (1980): J. Phys. C: Solid State Phys. **13**, 561 and L443

Lines, M.E., Glass, A.M. (1977): *Principles and Applications of Ferroelectrics and Related Materials*, The Int. Series of Monographs on Physics (Clarendon, Oxford)

Lischka, K. (1982): Appl. Physics A**29**, 177

Littlewood, P., Heine, V. (1979): J. Phys. C: Solid State Phys. **12**, 4431

Littlewood, P.B. (1979): J. Phys. C: Solid State Phys. **12**, 4441, 4459

Littlewood, P.B. (1980): J. Phys. C: Solid State Phys. **13**, 4855; **13**, 4875

Litvinov, V.I., Volkov, V.L., Tovstyuk, K.D. (1979) Ferroelectrics **22**, 839

Lopez-Otero, A. (1978): Thin Solid Films **49**, 1

Loudon, R. (1964): Adv. Phys. **13**, 423

Lowney, J.R., Senturia, S.D. (1976): J. Appl. Phys. **47**, 1771

Lucovsky, G., Martin, R.M., Burstein, E. (1971): Phys. Rev. B**4**, 1367
Lucovsky, G., White, R.M. (1973): Phys. Rev. B**8**, 660
Lyddane, R.H., Sachs, R.G., Teller, E. (1941): Phys. Rev. **59**, 673
Martinez, G., Schlüter, M., Cohen, M.I. (1975): Phys. Rev. B**11**, 651
Migoni, R., Bilz, H., Bäuerle, D. (1976): Phys. Rev. Lett. **37**, 1155
Mitter, H. (1981): *Ferroelektrischer Phasenübergang von $Pb_{1-x}Ge_xTe$*, Diploma Thesis, Johannes-Kepler-Universität Linz, Austria, unpublished
Muldawer, L. (1973): J. Nonmetals **1**, 177
Müller, K.A., Burkard, H. (1979): Phys. Rev. B**19**, 3593
Murase, K., Sugai, S., Takaoka, S., Katayama, S. (1976): Proc. 13th Conf. Phys. of Semiconductors, ed. by F.G. Fumi (Tipografica Marves, Rome) p. 305
Murase, K., Sugai, S. Higuchi, T., Takaoka, S., Fukunaga T., Kawamura, H. (1979): Proc. 14th Int. Conf. Phys. of Semiconductors, ed. by B.L.H. Wilson (The Institute of Physics, Bristol and London) p. 437
Murase, K., Sugai, S. (1979): Solid State Commun. **32**, 89
Murase, K. (1981): Ferroelectrics **35**, 67
Murase, K., Nishi, S. (1982): *Physics of Narrow-Gap Semiconductors*, Lecture Notes in Physics, Vol. 152, ed. by E. Gornik, H. Heinrich, L. Palmetshofer (Springer, Berlin, Heidelberg, New York) p. 261
Mycielski, J., Mycielski, A. (1978): Phys. Rev. B**18**, 1859
Natori, A. (1976): J. Phys. Soc. Japan **40**, 163; **41**, 782
Nii, R. (1964): J. Phys. Soc. Japan **19**, 58
Nill, K.W., Walpole, J.N., Calawa, A.R., Harman, T.C. (1971): In *The Physics of Semimetals and Narrow Gap Semiconductors*, ed. by D.L. Carter and R.T. Bate (Pergamon, Oxford) p. 383
Nishi, S., Kawamura, H., Murase, K. (1980): Phys. Status Solidi (b) **97**, 581
Oppermann, R., Thomas, H. (1975): Z. Phys. B**22**, 387
Ota, Y., Rabii, S. (1971): In *The Physics of Semimetals and Narrow-Gap Semiconductors*, ed. by D.L. Carter and R.T. Bate (Pergamon, Oxford) p. 343
Palik, E.D., Furdyna, J.K. (1970): Rep. Prog. Phys. **33**, 1195
Palmetshofer, L., Gresslehner, K.H., Ratschbacher, L., Lopez-Otero, A. (1982): In *Physics of Narrow-Gap Semicondcutors*, Lecture Notes in Physics, Vol. 152, ed. by E. Gorsik, H. Heinrich, L. Palmetshofer (Springer, Berlin, Heidelberg, New York) p. 391
Pawley, G.S., Cochran, W., Cowley, R.A., Dolling, G. (1966): Phys. Rev. Lett. **17**, 753
Pawley, G.S. (1968): J. Phys. (Paris), Colloque C4, Suppl. au no. 11-12, Thome **29**, C4-145
Penn, D.R. (1962): Phys. Rev. **128**, 2093
Perkowitz, S. (1969): Phys. Rev. **182**, 182
Perkowitz, S. (1975): Phys. Rev. B**12**, 3210
Perluzzo, G., Destry, J. (1978): Can. J. Phys. **56**, 453
Pine, A.S., Dresselhaus, G. (1971): Phys. Rev. B**4**, 356
Porod, W., Vogl. P., Bauer, G. (1980): Proc. 15th Int. Conf. Phys. of Semiconductors, Kyoto, ed. by S. Tanaka, Y. Toyozawa (Physical Society Japan, Tokyo): J. Phys. Soc. Japan **49**, Suppl. A, p. 649;
Porod, W., Vogl, P. (1982): *Physics of Narrow-Gap Semiconductors*, Lecture Notes in Physics, Vol. 152, ed. by E. Gornik, H. Heinrich, L. Palmetshofer (Springer, Berlin, Heidelberg, New York) p. 247
Pouget, J.P., Shapiro, S.M., Nassau, K. (1979): J. Phys. Chem. Solids **40**, 267
Pratt, G.W. (1973): J. Nonmetals **1**, 103; *Physics of IV-VI Compounds and Alloys*, ed. by S. Rabii (Gordon and Breach, London 1974) p. 85
Preier, H. (1979): Appl. Phys. **20**, 189
Rabii, S. (ed.) (1974): *Physics of IV-VI Compounds and Alloys*, Gordon and Breach, London
Rauluskiewicz, J., Górska, M., Kaczmarek, E. (eds.) (1978): *Physics of Narrow-Gap Semiconductors*, Proc. III. Int. Conf. of Narrow-Gap Semiconductors (PWN-Polish Scientific Publishers, Warsaw)
Ravich, Yu.I., Efimova, B.A., Smirnow, I.A. (1970): *Semiconducting Lead Chalcogenides*, ed. by L.S. Stil'bans (Plenum, New York)
Ravich, Yu.I., Efimova, B.A., Tamarchenko, V.I. (1971): Phys. Status Solidi (b) **43**, 11; **43**, 453
Rehwald, W., Lang, G.K. (1975): J. Phys. C: Solid State Phys. **8**, 3287
Richter, W. (1973): Private communication
Riedl, H.R., Dixon, J.R., Schoolar, R.B. (1967): Phys. Rev. **162**, 692

Rytz, D., Höchli, U.T., Bilz, H. (1980): Phys. Rev. B22, 359
Sawada, Y., Burstein, E., Carter, D.L., Testardi, L. (1965): Proc. of the Symposium on Plasma Effects in Solids, Paris 1964, ed. by J. Bok (Academic, New York) p. 71
Schmidt, H., Schwabl, F. (1978): Proc. Int. Conf. on Lattice Dynamics, Paris, ed. by M. Balkanski (Flammarion Sciences) p. 748
Schmidt, H., Schwabl, F. (1977): Phys. Lett. 61A, 476
Schneider, T., Beck, H., Stoll, E. (1976): Phys. Rev. B13, 1123
Schubert, K., Fricke, H. (1951): Z. Naturforschung 6a, 781; (1953) Structure Rep. 15, 72; Z. Metallkunde 44, 457; (1953) Structure Repts. 17, 44
Schwabl, F. (1972): Phys. Rev. Lett. 28, 500
Seddon, T., Farley, J.M., Saunders, G.A. (1975): Solid States Commun. 17, 55
Seddon, T., Gupta, S.C., Saunders, G.A. (1976): Phys. Lett. 56A, 45
Servoin, J.L., Gervais, F., Quittet, A.M., Luspin, Y. (1980): Phys. Rev. B21, 2038
Shimada, T., Kobayashi, K.L.I., Katayama, Y., Komatsubara, K.F. (1977): Phys. Rev. Lett. 39, 143
Shukla, G.C., Sinha, K.P. (1966): J. Phys. Chem. Solids 27, 1837
Snykers, M., Delavignette, P., Amelinchx, S. (1972): Mater. Res. Bull. 7, 831
Steigmeier, E.F., Harbeke, G. (1970): Solid State Commun. 8, 1275
Steigmeier, E.F., Harbeke, G., Wehner, R.K. (1971): In Structural Phase Transitions and Soft Modes, ed. by E.J. Gennielsen et al. (Universitetsforlaget Oslo) p. 409
Steigmeier, E.F., Auderset, H., Harbeke, G. (1975): Phys. Status Solidi (b) 70, 705
Stiles, P.J., Burstein, E., Langenberg, D.N. (1962): Phys. Rev. Lett. 9, 257
Sugai, S., Murase, K., Kawamura, H. (1977a): Solid State Commun. 23, 127
Sugai, S., Murase, K., Katayama, S., Takaoka, S., Nishi, S., Kawamura, H. (1977b): Solid State Commun. 24, 407
Sugai, S., Murase, K., Tsuchihira, T., Kawamura, H. (1979): J. Phys. Soc. Japan 47, 539
Sugimoto, N., Matsuda, T., Hatta, I. (1981): J. Phys. Soc. Japan 50, 1555
Suski, T., Bai, M., Zuczkowski, W., Kobayashi, K.L.I., Komatsubara, K.F. (1979): Solid State Commun. 30, 77
Suski, T., Dmowski, L., Baj, M. (1981): Solid State Commun. 38, 59;
    Suski, T., Baj, M., Katayama, S., Murase, K. (1982): In Physics of Narrow-Gap Semiconductors, Lecture Notes in Physics, Vol. 152, ed. by E. Gorsik, H. Heinrich, L. Palmetshofer (Springer, Berlin, Heidelberg, New York) p. 266
Suski, T., Kończykowski, M., Leszczyński, M., Lesueur, D., Dural, J. (1982): J. Phys. C: Solid State Phys. 15, L953
Sze, S.M. (1969): Physics of Semiconductor Devices (Wiley-Interscience, New York)
Szigeti, B. (1949): Trans. Faraday Soc. 45, 155
Takano, S., Hotta, S., Kawamura, H., Kato, Y., Kobayashi, K.L.I., Komatsubara, K.F. (1974): J. Phys. Soc. Japan 37, 1007
Takaoka, S., Murase, K. (1979): Phys. Rev. B20, 2823
Takasaki, K., Tanaka, S. (1977): Phys. Status Solidi (a) 40, 173 (1977)
Tennant, W.E., Cape, J.A. (1976): Phys. Rev. B13, 2450
Unterleitner, F. (1977): Untersuchung des paraelektrisch-ferroelektrischen Phasenübergangs an PbTe und $Pb_{1-x}Ge_xTe$, Diploma Thesis, Johannes-Kepler-Universität Linz, Austria, unpublished
Valassiades, O., Economou, N.A. (1975): Phys. Status Solidi (a) 30, 187
Vodop'yanov, L.K., Kucherenko, I.V., Shotov, A.P., Scherm, R. (1978): Pis'ma Zh. Eksp. Teor. Fiz. 27, 101 [English transl.: JETP Lett. 27, 92 (1978)]
Vogl, P., Vergês, J.A. (1982): Verhandlungen d. Deutschen Phys. Ges. 5/1982, p. 786
Volkov, V.L., Litvinov, V.I., Baginskii, V.M., Tovstyuk, K.D. (1976): Solid State Commun. 20, 807
Volkov, V.L., Litvinov, V.I. (1980): Phys. Lett. 75A, 398
Wagner, D., Bäuerle, D., Schwabl, F., Dorner, B., Kraxenberger, H. (1980): Z. Phys. B37, 317
Wagner, D., Bäuerle, D. (1981): Phys. Lett. 83A, 347
Wallace, P.R. (1965): Can. J. Phys. 43, 2162
Walpole, J.N., Nill, K.W. (1971): J. Appl. Phys. 42, 5609
Yaraneri, H., Grassie, A.D.C., Loram, J.W. (1982): Physics of Narrow-Gap Semiconductors, Lecture Notes in Physics, Vol. 152, ed. by E. Gornik, H. Heinrich, L. Palmetshofer (Springer, Berlin, Heidelberg, New York) p. 270
Zemel, J.N., Jensen, J.D., Schoolar, R.B. (1965): Phys. Rev. 140, A330
Zitter, R.N. (1971): Surf. Sci. 28, 335

# Electronic and Dynamical Properties of IV–VI Compounds

By A. Bussmann-Holder, H. Bilz, and P. Vogl

## 1. Introduction

### 1.1 History *)

In 1920 Valasek discovered that Rochelle salt exhibits ferroelectric properties.
He recognized that the dielectric properties of this material were in many respects
analogous to the ferromagnetic properties of iron. A ferroelectric crystal undergoes
a structural phase transition from an unpolarized high-temperature paraelectric
phase to a polarized low-temperature ferroelectric phase at a certain critical temperature $T_c$, the Curie temperature. In the ferroelectric state the polarization
of the crystal can be reversed or reoriented by means of an applied electric field
to an equivalent state. The dependence of polarization reorientation on the electric field is described by a hysteresis loop, i.e. at a certain critical field
strength $E_c$, the coercive field, the polarization switches into the reoriented
equivalent state. Reversal of the field up to $-E_c$ returns the crystal into the
original polarization state. Similar to a ferromagnet the phase transition is accompanied by an extremely large dielectric and piezoelectric response at and near
to $T_c$.

In 1933 the first model of a ferroelectric substance was proposed by Kurchatow.
He assumed that the dipolar forces between the constituent water molecules in
Seignette salt lead to the spontaneous polarization of the crystals. The failure
of this theory became obvious with the discovery of a series of isomorphous crystals by Busch and Scherrer from 1935-1938 (Busch and Scherrer, 1935; Busch, 1938),
the KDP-type ferroelectrics. This family of ferroelectrics is piezoelectric above
$T_c$ and undergoes a phase transition to either a ferroelectric or an antiferroelectric state, exhibiting very marked dielectric anomalies at $T_c$. With the discovery
of these crystals the first applications as underwater sound transducers and sub-
marine detectors were made based on their high electromechanical coupling effi-
ciency.

Slater presented in 1941 the first microscopic model to explain the occurrence
of ferroelectricity in these compounds. Compared to Seignette salt the KDP structure exhibits very simple features. The phosphate groups of KDP and its isomorphs

---

*) For a more extended historical survey see Lines and Glass, 1977.

are connected via hydrogen bonds with each other. Above $T_c$ the hydrogen ions are centered in the midth of each bond. Below $T_c$ they shift pairwise towards the phosphate group inducing by means of this movement a spontaneous polarization. Slater (1941) assumed that the hydrogen atoms are sitting in a double well potential in which they can tunnel from one minimum to the other by means of an applied electric field thus reversing the direction of the spontaneous polarization. Though this theory fails to explain the large shift in $T_c$ on deuteration (Bantle, 1942) in its essential features it is even nowadays one of the basic concepts used.

Up to 1945 it was generally believed that the phenomenon of ferroelectricity occurs only in crystals which contain hydrogen bonds. With the discovery of $BaTiO_3$ by Wul and Goldman (1945, 1946) this theory had to be abandoned as a unique picture of ferroelectricity. Soon after the discovery of $BaTiO_3$ the whole class of perovskites, their ceramics and their alloys were discovered by several researchers as for instance: Matthias (1949), Matthias and Remeika (1949), Remeika et al. (1950). Because of their high stability, their good growing conditions and their high dielectric anomalies a large field of applications was opened. Simultaneously the very simple structure of $ABO_3$-compounds led to an enormous increase in theoretical research. Ginzburg (1945, 1949) and Devonshire (1949, 1951, 1954) were the first who used a type of Landau theory for ferroelectric phase transitions.

In spite of the fact that the Landau theory has proven to be a very useful tool to describe structural phase transitions it misses the microscopic insight into the driving mechanism of ferroelectricity because of its phenomenological nature. In the early fifties and sixties the first real microscopic theories were developed by Fröhlich (1949), Cochran (1960) and Anderson et al. (1960). Starting point for these theories and the soft mode concept was the fact that a crystal lattice is only stable if all its eigenfrequencies are real (Born und Huang, 1954). In 1960 Cochran and Anderson stated that in ionic crystals one of their polar lattice modes may become imaginary in the harmonic approximation which leads to a ferroelectric phase transition. The reasoning for this statement was based on the fact that the harmonic force constants consist of two contributions with opposite signs: long-range attractive forces, i.e. Coulomb interactions, and short-range repulsive forces. If both parts are equal in magnitude one of the polar modes becomes zero thus inducing a change in structure, where the low temperature phase is determined by the frozen-in soft mode displacements.

A stabilization of the system in the paraelectric phase is achieved via anharmonic interactions. The observable quasiharmonic frequencies are positive above $T_c$ and approach zero as the temperature reaches $T_c$. Below $T_c$ the frequencies are zero thus determining by their displacement pattern the ferroelectric structure.

Since 1960 a number of theorists have been concerned with the soft mode concept. Common to all theories was the fact that the temperature dependence of the soft mode could be described by

$$\omega_f^2 = a(T - T_c)^\gamma \tag{1.1}$$

where $\gamma = 1$ is called the mean-field critical exponent.

The ferroelectric properties of the IV-VI compounds have first been suggested by Cochran et al. in 1966. Basically the theoretical attempts to provide a microscopic picture for the driving mechanism of ferroelectricity in these crystals can be divided into three models: the anharmonic lattice model, the vibronic model and the polarizability model. In forthcoming sections these models will be described in more detail and compared to each other.

## 1.2 Landau Theory

The Landau theory is based on the assumption that the free energy of a crystal, which undergoes a structural phase transition, can be expanded in powers of an order parameter. In ferroelectrics the order parameter is represented by the polarization P, which yields for the free energy

$$F = F_0 + \frac{1}{2} a P^2 + \frac{1}{4} b P^4 + \frac{1}{6} c P^6 + \dots \quad , \tag{1.2}$$

where it is assumed that only the coefficient of the squared term "a" exhibits a temperature dependence according to $a = a_0 (T - T_c)$, where $T_c$ is the phase transition temperature. The equilibrium conditions of the system under consideration are given by the first derivative of the free energy with respect to the order parameter, the polarization respectively

$$\frac{dF}{dP} = 0 = aP + bP^3 + cP^5 + \dots = 0 \quad . \tag{1.3}$$

Obviously this exhibits the trivial solution P = 0 for the paraelectric case, while for $T \leq T_c$ $P = \pm P_s$, where $P_s$ is the spontaneous polarization of the ordered phase and a function of a, b and c. A stable configuration is achieved if the second derivative of the free energy with respect to the order parameter is positive, i.e.

$$\frac{d^2F}{dP^2} > 0 \quad , \tag{1.4}$$

$$\frac{d^2F}{dP^2} = a + 3b P_s^2 + 5c P_s^4 = \chi^{-1} \quad , \tag{1.5}$$

where $\chi^{-1}$ defines the inverse isothermal dielectric susceptibility. Whether the phase transition is of first or second order is determined by the sign of the coefficient b. If b is positive the system undergoes a second order phase transition, i.e.

$$P_s = \pm\sqrt{\left[-\frac{a_o}{b}(T - T_c)\right]} \approx (T - T_c)^{\frac{1}{2}} \tag{1.6}$$

goes continuously to zero when the temperature is approaching $T_c$. The free energy as a function of the order parameter of the system which undergoes a second order phase transition is shown in Fig. 1.1 for three different temperatures. For $T < T_c$ the free energy has two equivalent minima at $P = \pm P_s$. At $T = T_c$ the two minima coalesce into a single broad minimum around $P = 0$. In the paraelectric phase the free energy exhibits a single minimum at $P = 0$.

A first order phase transition is characterized by $b < 0$. This yields for the polarization in the low temperature phase:

$$P_s = \pm\left\{-\frac{b}{c}\left[1 \pm\sqrt{1 - \frac{a_o c}{b^2}(T - T_c)}\right]\right\}^{\frac{1}{2}} \ . \tag{1.7}$$

At $T = T_c$ the polarization has still a finite value and at a temperature $T_1$, different from $T_c$, $P_s$ becomes zero. For this phase transition the free energy as a function of P shows the same behaviour for the paraelectric and the ferroelectric regime as in the case of second order. But it looks different at $T = T_c$ where the system is characterized by three equivalent minima (Fig.1.2).

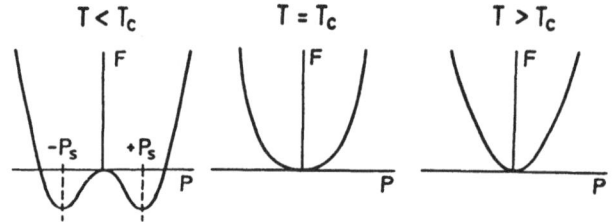

Fig. 1.1  Free energy as a function of polarization at various temperatures in case of a second-order phase transition

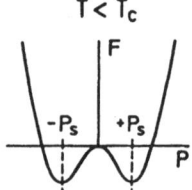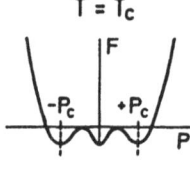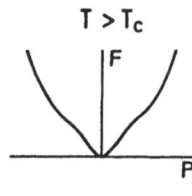

Fig. 1.2  Free energy as a function of polarization at various temperatures in case of a first-order phase transition

## 2. Chemical Structure and Electronic Bands of IV-VI Compounds

### 2.1 Structure and Ferroelectricity

The IV-VI compounds have 10 electrons per unit cell, such as the group V elements and the III-VII compounds. All of these systems crystallize either in the rock-salt structure, characteristic of ionic compounds, or in a distorted rock-salt structure where covalent bonds can be formed (see Table 2.1 and reviews by Ravich et al., 1970; Dalven, 1973; Lovett, 1977; Ley et al., 1979). GeTe and SnTe become ferro-

Table 2.1 Stable modification of group V, IV-VI and III-VII compounds. Unstable modifications can sometimes be induced by epitaxy (Mariano and Chopra, 1967). The polarity, given in the upper right corner of each corresponding box, is defined as half of the difference of the atomic p-orbital energies of cation and anion (see main text). In addition, the minimum band gap and the high frequency dielectric constant $\varepsilon_\infty$ is given; for the lead chalcogenides low temperature values are tabulated. The values for $\varepsilon_\infty$ in the orthorhombic structures correspond to the a, b, and c-axis, respectively

| Compound / Element | Polarity (eV) | Structure | Band gap (eV) | $\varepsilon_\infty$ |
|---|---|---|---|---|
| **Group V** | | | | |
| P (black) | — | □[*]  80<p<110 kbar u) | 0.35 a) | — |
| As | — | ◊[*] | semimetal | — |
| Sb | — | ◊[*] | semimetal | — |
| Bi | — | ◊[*] | semimetal | — |
| **IV – VI** | | | | |
| GeS | 2.14 | □  p<340 kbar v) | 1.65 b) | 13.5 13.0 10.0 c) |
| GeSe | 1.68 | □  p<340 kbar v) | 1.54 d) | 18.7 21.9 14.4 e) |
| GeTe | 1.11 | ◊  T>670 K f) | 0.49 g) | 36 h) |
| SnS | 2.42 | □  p<340 kbar v) / □ Epitaxial | 1.1 i) | 14 16 16 j) |
| SnSe | 1.96 | □  p<340 kbar v) / □ Epitaxial | 0.90 k) | 13 17 16 j) |
| SnTe | 1.39 | ◊  T<98 K f) / □ p>14 kbar l) | 0.3 m) | 45 h) |
| PbS | 2.54 | □ / □ p>25 kbar / □[*] p>215 kbar v) | 0.286 n) | 17.2 h) |
| PbSe | 2.08 | □ / □ p>43 kbar p) / □[*] p>160 kbar v) | 0.165 n) | 22.9 h) |
| PbTe | 1.51 | □ / □ p>55 kbar / □[*] p>130 kbar | 0.19 n) | 32.8 h) |
| **III – VII (Tl)** | | | | |
| TlF | 7.32 | □ | — | — |
| TlCl | 4.28 | □[*] / □ Epitaxial | 3.4 q) | 5.00 r) |
| TlBr | 3.60 | □[*] / □ Epitaxial | 3.0 q) | 5.34 r) |
| TlI | 2.87 | □[**]  T>440 K p>5 kbar s) / □[*] | 2.79 | 7.0 t) |

Legend (structure symbols):

- □[*]  $D_{2h}^{18}$  orthorhombic
- ◊[*]  $D_{3d}^{5}$  rhombohedral
- □  $D_{2h}^{16}$  orthorhombic
- ◊  $C_{3v}^{5}$  rhombohedral
- □  $O_{h}^{5}$  NaCl-structure
- □[**]  $D_{2h}^{17}$  orthorhombic
- □[*]  $O_{h}^{1}$  CsCl-structure

References:

a) D. Warschauer, 1963
b) Wiley et al., 1975
c) Wiley et al., 1976
d) Vlachos et al., 1976
e) Siapkas et al., 1976
f) Murase, 1980; Jantsch, this volume
g) Polatoglou et al., 1982
h) Burstein et al., 1971
i) Lambros et al., 1974
j) Chandrasekhar et al., 1977
k) Albers et al., 1962
l) Suski et al., 1981
m) Moldovonova et al., 1966
n) Mitchel et al., 1964
p) Mariano and Chopra, 1967
q) Heidrich et al., 1974
r) Lowndes and Martin, 1969
s) Heidrich et al., 1975
t) Levine, 1973
u) Doggin, 1972
v) Chattopadhyay et al., 1983

electric at $T_c$ = 670 K and $T_c$ = 98 K, respectively (Kawamura, 1977; Murase, 1980; Katayama and Mills, 1980; Jantsch, this volume; Schubert and Fricke, 1951, 1953; Steigmeier and Harbeke, 1970); they show the same 3-fold coordinated rhombohedral structure as the group V elements. This structure may be obtained from the NaCl structure by displacing the two sublattices relative to one another along the <111> axis and by stretching the unit cell along the <111> axis (Cowley, 1968; Goldak et al., 1966; Hohnke et al., 1972; Schubert and Fricke, 1951, 1953). The other Ge and Sn IV-VI compounds crystallize in the orthorhombic black phosphorus structure which also has a 3-fold covalent coordination. This structure can be derived from the NaCl structure by displacing the two sublattices relative to one another in the <100> direction of the orthorhombic cell, with the distortion reversing for every other atomic plane (Pawley, 1968; Shalvoy et al., 1977). It was pointed out by Pawley (1968) that this distortion provides an antiferroelectric nature for the orthorhombic materials. The rocksalt structure compounds PbS, PbSe, and PbTe, can be transformed into this structure by hydrostatic pressure (see Table 2.1). This may be expected from the fact that high pressure favours covalency (Schiferl, 1974; Kemeny et al., 1977; Ley et al., 1979).


## 2.2 Ionicity and Covalency

The crystal binding mechanisms and crystalline phases have been investigated theoretically quite extensively, using various types of phenomenological approaches, such as ionicity concepts (Stiles and Brodsky, 1972; Schiferl, 1974; Shalvoy et al., 1977) quantum chemistry methods (Tanaka and Morita, 1979), or ideas related to pseudopotentials and ionic radii (Littlewood, 1980, 1982; Mula, 1982). Since the IV-VI compounds are on the borderline between ionic and covalent crystals, it is not surprising that the various proposed ionicity scales differ quite appreciably from one another (Kowalczyk et al., 1974; Schiferl, 1974; Shalvoy et al., 1977). The simplest and perhaps most plausible scheme is based on the "Atomic Electronegativity Scale", proposed by Harrison (1980) and applied to IV-VI compounds by Nakaniski and Matsubara (1980). These authors define a "polarity" $V_3 = 0.5 (E_p^{cation} - E_p^{anion})$, given by half of the difference between the free atomic (outermost) p-orbital energies of the constituent cation and anion atoms in the compound. This electronegativity scale is quite similar to Pauling's (1960) ionicity. We have included $V_3$ in Table 2.1; the orbital energies are taken from published atomic Hartree-Fock calculations (Fischer, 1972). This quantity $V_3$ comes from a tight-binding analysis of IV-VI compounds. The charge transfer from cation to anion in the solid is proportional to $V_3$, and the average energy gap is, in a simple tight-binding model, given by $2V_3$ (Harrison, 1980). This value qualitatively agrees with the average dielectric gap (Kawamura, 1977). As can be further deduced from Table 2.1, the trends in $V_3$ are in accord with the observed trends in the ferroelectric phase transition temperatures in the series GeTe, SnTe, PbTe.

## 2.3 Electronic Band Structure

The band structure of IV-VI compounds has been investigated theoretically and experimentally quite extensively. Quantitative band structure calculations have been performed for PbTe (Tung and Cohen, 1969; Herman et al., 1968; Bernick and Kleinman, 1970; Rabii and Lasseter, 1974; Martinez et al., 1975), PbSe (Herman et al., 1968; Bernick and Kleinman, 1970; Kohn et al., 1973; Rabii and Lasseter, 1974; Martinez et al., 1975), PbS (Herman et al., 1968; Bernick and Kleinman, 1970; Kohn et al., 1973; Rabii and Lasseter, 1974), SnTe (Herman et al., 1968; Tung and Cohen, 1969; Bernick and Kleinman, 1970; Melvin and Hendry, 1969), SnSe (Abbati et al., 1977; Car et al., 1978), SnS (Parke and Srivastava, 1980), GeTe (Herman et al., 1968; Tung and Cohen, 1969; Polatoglou et al., 1982), and GeS (Grandke and Ley, 1977).

Experimentally, the electronic structure of these materials was investigated by angular resolved photoemission measurements (Grandke and Ley, 1977; Grandke et al., 1978; Ley et al., 1979), angular integrated photoemission or x-ray emission studies (Kemeny et al., 1977; Shevchik et al., 1973; Kemeny and Cardona, 1976), or by optical absorption, reflectivity and Raman measurements (Mitchell et al., 1964; Burstein et al., 1971; Vlachos et al., 1976; Siapkas et al., 1976; Chandrasekhar and Zwick, 1976; Wiley et al., 1975, 1976; Albers et al., 1962; Moldovonova et al., 1966).

A typical example for the band structure of a IV-VI compounds is depicted in Fig. 2.1. It shows the band structure and the density of states for PbTe, as obtained from the relativistic empirical pseudopotential calculation of Martinez et al. (1975).

The lower part of the valence band consists of the non-bonding Te (5s) states and, above them, the Pb (6s) states which are somewhat hybridized with the Te (5p) states. The upper part of the valence band and the lower part of the conduction band consist mainly of Te (5p) and Pb (6p) states, respectively. The band gap is direct but occurs at the L point. It is very unusual in semiconductors that the valence bands reach their maximum away from $\vec{k}$ = 0. This is a consequence of the large spin-orbit interaction associated with the heavy Te (and also Pb) atoms and is common to all lead chalcogenides and to SnTe and GeTe (Grandke et al., 1978). The presence of the (6s)-levels of Pb immediately below the (5p)-levels of Te lifts up the L valence bands as a result of the s-p admixture which is forbidden at $\vec{k}$=0 (Ley et al., 1979). In PbTe, the conduction band minimum is formed by the odd $L_6^-$ state, and the valence band edge has the $L_6^+$ symmetry. The same ordering occurs in PbTe and PbS, while in SnTe it is reversed: $L_6^-$ forms the valence band edge and

57

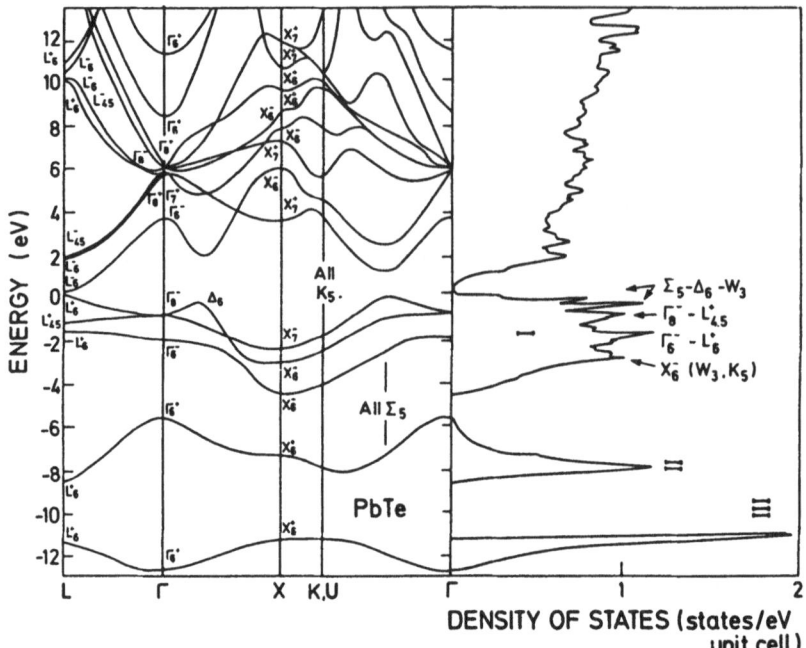

**Fig. 2.1** Pseudopotential band structure of PbTe, together with the calculated density of states. From Martinez et al. (1975)

$L_6^+$ the conduction band edge. This has the interesting consequence that the band gap of the alloy $Pb_{1-x}Sn_xTe$ vanishes for a certain alloy concentration, $x \approx 0.4$ (see next section).

In Fig. 2.2 we show the variation of the direct band gap through the series SnTe-PbTe-PbS-PbSe-SnSe, including its variation in the binary alloys formed with these compounds (Preier, 1979). Both the temperature and the pressure dependence of the direct gap is anomalous in the IV-VI compounds when compared to most other semiconductors. The direct energy gap increases with temperature and decreases with pressure (Dalven, 1973). The temperature dependence of the gap is shown in Fig.2.3 for the lead salts. Above approximately 75 K the band gap is a linear function of temperature and becomes nonlinear below 75 K. The temperature coefficient is remarkably universal in all compounds as shown in Fig. 2.3, namely $(\partial E_g/\partial T)_p \approx$

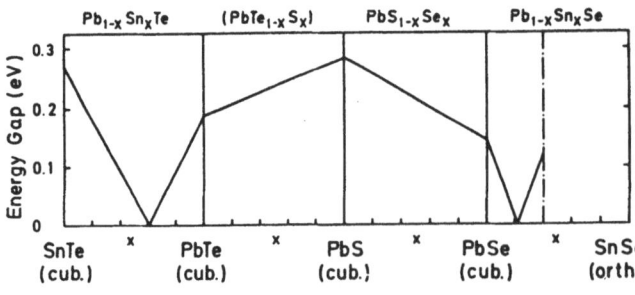

**Fig. 2.2** Variation of the direct energy gap of various IV-VI mixed crystals as a function of the alloy concentration. After Preier (1979)

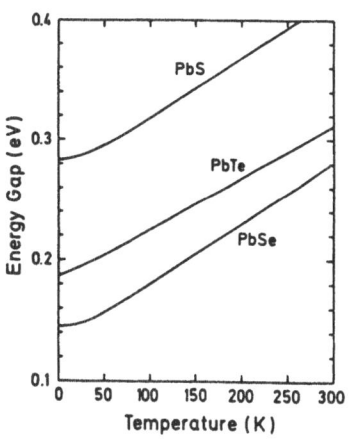

Fig. 2.3 Temperature variation of the direct energy gaps of the lead salts. After Preier (1979)

$4.7 \cdot 10^{-4}$ eV/K. The pressure coefficient of the gap is also almost the same in all IV-VI compounds, and is approximately given by $(\partial E_g/\partial P)_T \approx -9 \cdot 10^{-6}$ eV/bar (Dalven, 1973).

## 2.4 The Zero-Gap Situation

As pointed out before, the tin chalcogenides SnTe and SnSe have their gap close to the L point but the symmetry of valence and conduction band is reversed compared to PbTe or PbSe. Since the energy gap in the alloys is intermediate between the values of the pure compounds one expects a zero-gap situation for the alloys $Sn_{1-x}Pb_xTe$ and $Sn_{1-x}Pb_xSe$ (see Fig.2.2). The zero-gap situation has been confirmed experimentally for both alloys by measurements of the electrical resistivity, the Hall coefficient and optical measurements of the gap. The position of the energy bands as a function of composition x is shown in more detail in Fig.2.4 for $Pb_{1-x}Sn_xTe$. The direction of change by the band structure under the effects of temperature, pressure and magnetic field are shown in the same Fig.2.4. The effect

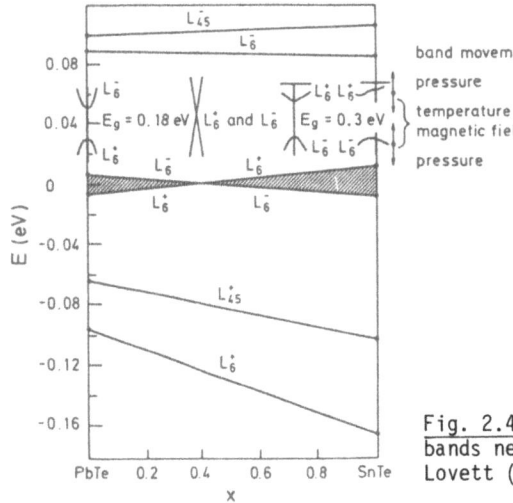

Fig. 2.4. Approximate positions of the energy bands near the L-point for $Pb_{1-x}Sn_xTe$. After Lovett (1977)

59

of the zero-gap on the ferroelectricity of IV-VI alloys was investigated theoretically and experimentally by Kawamura (1977), Murase (1980), Katayama and Mills
(1980), and particularly by Nishi et al. (1980). The latter authors measured the
static dielectric constant in $Pb_{1-x}Sn_xTe$ and could not detect any effect of the
band inversion on the dielectric constant. This finding is in contrast to previous
experiments by Kobayashi et al. (1979) where an appreciable change in the transition temperature with the minimum gap was reported. From the theoretical side, no
such dependence is expected: Semiempirical models (Murase et al., 1979; Littlewood,
1980, 1982) and pseudopotential calculations (Porod and Vogl, 1982; see Sect. 3.2)
predict the soft mode and the stability of the cubic structure to be insensitive
to the band edge structure, basically because of the small electronic density of
states close to the energy gap.

## 3. Lattice Dynamics and Phase Transitions

### 3.1 General Aspects: The Soft-Mode Concept

In a simple diatomic cubic ionic solid the two optic frequencies at long wavelengths
are connected with the two dielectric constants by the Lyddane-Sachs-Teller relation:

$$\frac{\omega^2_{TO}}{\omega^2_{LO}} = \frac{\varepsilon_\infty}{\varepsilon_0} \qquad (3.1)$$

Here the subindices denote: TO = transverse optic, LO = longitudinal optic, $\infty \triangleq$
'high-frequency' regime (above the infrared regime), and o $\triangleq$ static value. Since
$\varepsilon_0^{-1} \to \infty$ as $T \to T_c$ in a displacive ferroelectric phase transition, and $\omega_{LO}$ as well
as $\varepsilon_\infty$ change very little only in this transition, it follows that

$$\omega^2_{TO} \to 0 \text{ with } T \to T_c \quad . \qquad (3.2)$$

This observation by Fröhlich (1949) was the starting point for the later development of the soft-mode concept by Cochran (1960) and Anderson (1960) in which the
lowest transverse optic mode, $\omega^I_{TO} = \omega_F$, (in a general ferroelectric like $KNbO_3$),
may be related to the transition temperature $T_c$ by the mean-field relation (see
Sect. 3.3)

$$\omega^2_f = A (T - T_c)^\gamma , \gamma = 1 \quad . \qquad (3.3)$$

Generally, $\gamma$ deviates from the value one and may change from the quantum limit
$\gamma = 2$ if $T_c = 0$ to $\gamma \approx 1/3$ above the mean-field regime.
The different miscroscopic theories and models will be discussed in the subsequent sections. They are all based on the critical cancellation of long- and
short-range forces in the ferroelectric phase transition first discussed by Slater

(1941). The ionicity, the covalency and the polarity of the ferroelectric materials exhibit an interesting playing together in the phase transition. At present, three different models are discussed in the literature giving different descriptions of the 'driving mechanism' which governs the temperature dependence of the soft mode. The 'pro's' and 'contra's' of these models will be presented in the following.


## 3.2 Electronic Theory of the Soft-Mode Instability

Several first-principle (Yin and Cohen, 1982a, 1982b; Kunc and Martin, 1982) and parameter-free approaches (Boyer, 1981) have been developed recently which may be used to investigate ferroelectricity from a microscopic point of view in terms of the electronic structure.

The pseudopotential method has been used recently by Porod et al. (1980) and by Porod and Vogl (1982) to develop a simple model for zone center phonon modes and ferroelectricity. Since this model was applied specifically to IV-VI compounds (Porod and Vogl, 1982), we shall review it in some detail now.

Following Porod and Vogl (1982), we shall calculate in this section the harmonic part of the zone-center transverse-optical phonon mode $\omega_{TO}$ for cubic, binary crystals in terms of pseudopotential theory. It was pointed out in Sect. 3.1 that $\omega_{TO}$ is a characteristic measure of the stability of a crystal against a ferroelectric distortion.

In pseudopotential theory (Cohen and Heine, 1970; Yin and Cohen, 1982a, 1982b) it is customary to divide the total crystal charge distribution into rigid, positively charged ion cores (with charges $Z_c = + 4$ and $Z_a = + 6$ for the cation Pb and anion Te in PbTe, for example) and all valence electrons. The force in x-direction exerted on an anion core at $L = (\ell a)$ ($\ell$ = lattice cell) due to the small x-displacement of a cation core at $L' = (\ell' c)$ consists of two parts:

(i) The force $\phi_{xx}^I(L, L')$ due to the Coulomb field of the displaced core $L'$.

(ii) The force $\phi_{xx}^E(L, L')$ due to the potential set up by the valence electrons which are redistributed by the displacement of $L'$. This gives (Sham, 1974)

$$\phi_{xx}^I(L,L') = Z_a Z_c \frac{\partial^2 v(\vec{x}_L - \vec{x}_{L'})}{\partial x_L \, \partial x_{L'}} \quad , \tag{3.4}$$

$$\phi_{xx}^E(L,L') = \int d^3r \, \frac{\partial v(\vec{r} - \vec{x}_L , a)}{\partial x_{L'}} \, \delta n(\vec{r},L') \quad , \tag{3.5}$$

$$\delta n(\vec{r},L') = \int d^3r' \, \chi(\vec{r},\vec{r}') \, \frac{\partial v(\vec{r}' - \vec{x}_{L'} , c)}{\partial x_{L'}} \quad . \tag{3.6}$$

61

In these equations, $v(\vec{r}) = e^2/r$, and $v(\vec{r}, a)$ is the bare anion electron-ion pseudopotential which is Coulombic outside the core and modified inside (Cohen and Heine, 1970). Following Porod and Vogl (1982), we consider only local, non-relativistic pseudopotentials, although this is a rather crude approximation for IV-VI compounds. In Eq. (3.6), $\chi(\vec{r},\vec{r}')$ is the static (valence) electron density response function of the perfect crystal. It is related to the dielectric function $\varepsilon(\vec{r},\vec{r}')$ and the irreducible electron polarizability $\chi^P(\vec{r},\vec{r}')$ by the matrix relations (Sham, 1974)

$$\varepsilon^{-1} = 1 + v\chi \quad , \varepsilon = 1 - v\chi^P \tag{3.7}$$

where v is the Coulomb potential. The eigenfrequencies can be obtained by Fourier transforming the force constants, Eqs. (3.4)-(3.6). In Fourier space the response functions read $\chi(\vec{q}+\vec{G}, \vec{q}+\vec{G}')$ with $\vec{G}, \vec{G}'$ reciprocal lattice vectors and the wave vector $q \in 1.BZ$. As shown in detail by Sham (1974), one obtains from Eqs. (3.4)-(3.6) for cubic, binary crystals

$$\omega_{TO}^2 = \frac{\hbar^2 e^2}{M_{red} \Omega_0} (I - E), \tag{3.8}$$

$$I = \frac{4\pi}{3} Z_a Z_c \quad , \tag{3.9}$$

$$E = \frac{\Omega_0^2}{4e^2} \sum_{\vec{G},\vec{G}' \neq 0} \delta v_x(\vec{G},a) \, \chi(\vec{G},\vec{G}') \, \delta v_x(\vec{G}',c) \quad , \tag{3.10}$$

$$\delta v_x(\vec{G},a) = G_x \, v(\vec{G},a) \, e^{-i\vec{G}\cdot\vec{x}_a} \quad . \tag{3.11}$$

$M_{red}$ and $\Omega_0$ are the reduced unit cell mass and volume, respectively.

The term I in Eq. (3.9) originates in the short-range $\vec{q} = 0$ part of the ion-ion interaction $\phi_{xx}^I$, Eq. (3.4), and is obtained through Ewald's method (see Sham, 1974). It simply describes the plasma vibration of the positive ion cores embedded in an homogeneous negative background. Similarly, the electronic part E originates in $\phi_{xx}^E$, Eq. (3.5). This term tends to reduce or "screen" the plasma vibration frequency of the ion cores. Typically, one finds $I \simeq 70$ and $E \simeq 30$ in covalent semiconductors such as Si or GaAs, while $(I - E)/I < 10^{-3}$ in IV-VI compounds (Porod et al. 1980).

The expression Eq. (3.11) represents the bare electron-ion potential induced by a small unit x-displacement of sublattice a. It was pointed out by Porod et al. (1980, 1982) that Eq. (3.10) can be rewritten such as to contain the induced *total*, screened electron-ion potential, given by

$$\delta V_x(\vec{G},a) = \sum_{\vec{G} \neq 0} \varepsilon^{-1}(\vec{G},\vec{G}') \, \delta v_x(\vec{G}',a) \quad , \tag{3.12}$$

rather than the bare potential, and the dielectric function $\varepsilon(\vec{G},\vec{G}')$ itself rather than its inverse [which corresponds to $\chi(\vec{G},\vec{G}')$]. This considerably simplifies the self-consistent calculation of Eq. (3.10). So far the analysis is quite general and rigorous.

With two approximations one can reduce Eqs. (3.8)-(3.12) to simple analytical expressions which can be evaluated by hand (Porod et al., 1980, 1982). For Eq. (3.12) the rigid-ion approximation is used and the off-diagonal elements of $\varepsilon(\vec{G},\vec{G}')$ for $\vec{G}$ and $\vec{G}' \neq 0$ are neglected. The electronic part of the force constants, E in Eq. (3.10), then reduces to

$$E = \frac{\Omega_0^2}{4e^2} \sum_{\vec{G}\neq 0} e^{-i\vec{G}\cdot\vec{x}_a} G_x V(G,a)\varepsilon_G \chi(\vec{G},\vec{G})\varepsilon_G G_x V(G,c)e^{i\vec{G}\cdot\vec{x}_c}$$

$$= \frac{\Omega_0^2}{12e^2} \sum_{\vec{G}\neq 0} S(G)G^2[V_S^2(G)-V_A^2(G)](G^2/4\pi e^2)\varepsilon_G(1-\varepsilon_G) \ , \tag{3.13}$$

$$S(G) = \sum_{\text{star of } \vec{G}} \exp[i\vec{G}\cdot(\vec{x}_c - \vec{x}_a)] \quad . \tag{3.14}$$

In these equations, $V(G,a)$ and $V(G,c)$ denote the screened anion and cation perfect crystal pseudopotential form factors and $\varepsilon_G \equiv \varepsilon(\vec{G},\vec{G})$. In Eq. (3.13), $V_{S,A}(G) = 0.5[V(G,a) \pm V(G,c)]$ are the symmetric and antisymmetric form factors. One can further simplify Eq. (3.13) by noting that the semiconductor dielectric function $\varepsilon_G$ differs only by a few percent from the free electron gas value for $G \neq 0$. Since $\varepsilon_G$ for free electrons is known (Cohen and Heine, 1970), and $V_S(G)$ and $V_A(G)$ are tabulated in the literature (Cohen and Heine, 1970; Animalu et al., 1966), and the G-sum in Eq. (3.13) is rapidly convergent, $\omega_{TO}^2$ can easily be calculated from Eqs. (3.8), (3.9), (3.13), (3.14) for any cubic, binary solid. As shown by Porod et al. (1980) and Porod and Vogl (1982), the resulting TO frequencies agree very well with the experimental data for covalent semiconductors, but are only semi-quantitative for ionic solids and IV-VI's due to the neglect of relativistic effects and non-locality in the pseudopotentials. For PbTe, for example, it is found (Porod and Vogl, 1982) that E(theory) = 82.72 while E(experimental) = 100.09.

The major advantage of this approach, however, is to clearly demonstrate the physical origin of ferroelectricity or soft TO-phonon modes. $\omega_{TO}^2 \simeq 0$ requires a large electron-mediated part of the force constants in Eq. (3.8), such that $E \simeq I$. This requires the following two properties of the crystal:

(i) Ionic structure, i.e. NaCl or CsCl structure.

(ii) Covalency, in the sense that $V_S(G)$ is large compared to $V_A(G)$ for the leading $\vec{G}$'s.

The second condition is a prerequisite for getting a large factor $V_S^2 - V_A^2$ in Eq. (3.13), Ionic solids like NaCl do not obey this condition (ii); since $V_{Cl} \gg V_{Na}$

one gets $V_S \simeq V_A$ for the leading $\vec{G}$'s ($V_A(111)/V_S(111) = 0.7$ in NaCl (Cohen and Heine, 1970)). Ionic solids which do not obey condition (ii) therefore have a stable lattice.

The first condition originates in the geometry-dependent structure factor $S(G)$, Eq. (3.14). In zincblende or diamond lattices, $S(\vec{G}) = 0$ for the first star $\vec{G} \in \{111\}$, where $V_S(G)$ is largest. Since $V_S(G)$ decreases steeply with increasing G, E is small in this case. This implies that the covalent semiconductors such as Si or GaAs, which do obey condition (ii) but not (i), are also stable with respect to ferroelectric distortions.

The IV-VI compounds, on the other hand, fulfill both conditions (i) and (ii). For this class of materials, the theory predicts soft TO phonon modes, in agreement with experiment. In PbTe, for example, $V_A(111)/V_S(111) = 0.2$ (Animalu et al., 1966) which implies a "covalency" comparable to the III-V compounds. The III-VII thallous halides represent another class of materials which fulfill both conditions (i) and (ii) and are indeed known to have soft, i.e. strongly temperature dependent TO phonon frequencies (Lowndes, 1972).

One may summarize these findings as follows. A crystal can be expected to show soft TO modes if it has an ionic structure, while the weak electronegativity of its constituent ions favours already covalent bonding. Ionic structure and covalency then together produce the ferroelectric instability. We point out, however, that neither the direct energy gap nor the macroscopic dielectric constant $\varepsilon_\infty$ of the material play a role for $\omega_{TO}^2$.

We conclude that this investigation clarifies the covalent interaction to be a decisive factor for ferroelectricity in the highly symmetric IV-VI compounds. In many of the "classic" ferroelectrics, such as the perovskites, the strong anisotropy of the local field is also known to be essential (Slater, 1950). Further investigations are needed to clarify the quantitative role of covalency in these materials; doubtless it is a supporting element for the ferroelectric instability (Bussmann et al., 1981).

So far, the existence of a soft-mode instability does not give a clue to the 'driving' mechanism which is responsible for the para - ferroelectric phase transition. In contrast to the situation in superconductivity where the electron-phonon coupling has been accepted since more than two decades as being the microscopic mechanism (at least for the overwhelming majority of cases) the situation in ferroelectricity is still controversial. The different microscopic models which are discussed, at the present time, will be reviewed in the following section.

## 3.3 The Anharmonic Lattice Model

### 3.3.1 Quasi-Harmonic Approximation

It has been pointed out in the previous section that a crystal lattice is stable against small deformations if all its normal phonon modes have real frequencies. Cochran (1960) and Anderson (1960) showed that in some ionic crystals one of the polar lattice modes may become imaginary in the harmonic approximation because of the cancellation of short-range repulsive and long-range attractive forces. A stabilization of the system above the phase transition temperature can be achieved via anharmonic interactions.

The Hamiltonian of an anharmonic crystal within the renormalized-phonon approximation (Thomas, 1969) can be written as

$$H = \frac{1}{2} \sum_{\lambda,q} M_\lambda \dot{Q}^2(\lambda,q) + \frac{1}{2} \sum_{\lambda,q} \omega_o^2(\lambda,q)\, Q(\lambda,q)\, Q(\lambda,-q)$$

$$+ \frac{1}{4} \sum_{\substack{\lambda\lambda'\mu\mu' \\ qq'\kappa\kappa'}} V^{(4)}_{\lambda\lambda'\mu\mu'}(qq',\kappa\kappa')\, Q(\lambda,q)\, Q(\lambda',q')\, Q(\mu,\kappa)\, Q(\mu',\kappa') \quad , \tag{3.15}$$

where $Q(\lambda,q)$ are the normal mode coordinates and $\omega_o^2(\lambda,q)$ are their respective frequencies squared which are positive in a stable crystal lattice (Born and Huang, 1954). An instability which leads to a phase transition requires that at least one of these normal mode frequencies becomes imaginary. Higher order anharmonic terms have not been taken into account in the above Hamiltonian for the sake of simplicity. The above written Hamiltonian can be approximated by an effective quadratic Hamiltonian by replacing the fourth-order term by a quadratic term, i.e. $V^{(4)}Q^4 = 6V^{(4)}Q^2 <Q^2>$ where $<Q^2>$ represents the thermal average over the normal mode displacements squared. The anharmonic Hamiltonian thus becomes quadratic

$$H = \frac{1}{2} \sum_{\lambda,q} M_\lambda \dot{Q}^2(\lambda,q) + \frac{1}{2} \sum_{\lambda,q} \omega_o^2(\lambda,q)\, Q(\lambda,q)\, Q(\lambda,-q)$$

$$+ \frac{1}{4} \sum_{\substack{\lambda\lambda'\mu\mu' \\ q\kappa}} \bar{V}^{(4)}_{\lambda\lambda'\mu\mu'}(q,\kappa) <Q(\mu,\kappa)\, Q(\mu',-\kappa)> Q(\lambda,q)\, Q(\lambda',-q) \quad , \tag{3.16}$$

where the coefficients $\bar{V}^{(4)}(q,\kappa)$ are expressed in terms of the quartic anharmonic force constants $V^{(4)}(q,\kappa)$. Because of the interaction between phonons the effective Hamiltonian is non-diagonal.

To get the new renormalized temperature dependent normal modes frequencies $\bar{\omega}(\lambda,q)$ in a self-consistent manner, the diagonalization is performed by means of a unitary transformation for $Q(\lambda,q)$

$$\tilde{Q}(\lambda,q) = \sum_\mu \alpha_{\lambda\mu}(q)\, Q(\mu,q) \quad , \tag{3.17}$$

$$\tilde{Q}(\lambda,-q) = \sum_\mu \alpha^*_{\lambda\mu}(q)\, Q(\mu,-q) \quad , \tag{3.18}$$

$$\sum_\nu \alpha_{\mu\nu}\alpha^*_{\mu'\nu} = \delta_{\mu\mu'} \quad , \tag{3.19}$$

and in an analogous way for the momenta $P(\lambda,q)$. The proper determination of the coefficients $\alpha$ leads to an effective transformed Hamiltonian in diagonal form:

$$H_{eff} = \frac{1}{2} \sum_{\lambda,q} \tilde{P}(\lambda,q)\,\tilde{P}(\lambda,-q) + \frac{1}{2} \sum_{\substack{q \\ \lambda\lambda'\nu}} \omega_0^2(\nu,q)\,\alpha^*_{\lambda\nu}(q)\alpha_{\lambda'\nu}(q)$$

$$\times \tilde{Q}(\lambda,q)\,\tilde{Q}(\lambda',-q) + \frac{1}{4} \sum_{\substack{q,\kappa \\ \lambda\lambda'\mu\mu'\nu\nu'}} \tilde{V}^{(4)}_{\nu\nu'\mu\mu'}(q,\kappa) <Q(\mu,\kappa)\,Q(\mu',-\kappa)>$$

$$\times \alpha_{\lambda\nu}(q)\alpha_{\lambda'\nu'}(q)\,\tilde{Q}(\lambda,q)\,\tilde{Q}(\lambda',-q) \quad , \tag{3.20}$$

from which the following condition for the coefficients results:

$$\frac{1}{2} \sum_\nu \omega_0^2(\nu,q)\,\alpha^*_{\lambda\nu}(q)\,\alpha_{\lambda'\nu}(q) + \frac{1}{4} \sum_{\substack{\kappa \\ \mu\mu'\nu\nu'}} \tilde{V}^{(4)}_{\nu\nu'\mu\mu'}(q,\kappa) <Q(\mu,\kappa)\,Q(\mu',-\kappa)>$$

$$\times \alpha^*_{\lambda\nu}(q)\,\alpha_{\lambda'\nu'}(q) = \frac{1}{2}\bar{\omega}^2(\lambda,q)\delta_{\lambda\lambda'} \quad . \tag{3.21}$$

This leads to an effective harmonic Hamiltonian in diagonal form with temperature dependent renormalized normal mode frequencies $\bar{\omega}(\lambda,q)$. The eigenstates of this Hamiltonian are harmonic oscillator states and the temperature dependence enters via the averages of $<Q(\mu,q)Q(\mu',-q)>$. In a first approximation, one can assume that the renormalized phonon eigenvectors are the same as in the harmonic case, i.e.

$$\tilde{Q}(\lambda,q) = Q(\lambda,q) \rightarrow \alpha_{\lambda\mu}(q) = \delta_{\lambda\mu} \quad , \tag{3.22}$$

only the phonon frequencies are renormalized:

$$\bar{\omega}^2(\lambda,q) = \omega_0^2(\lambda,q) + 2 \sum_{\kappa\mu\mu'} V^{(4)}_{\lambda\lambda\mu\mu'}(q,\kappa) <Q(\mu,\kappa)\,Q(\mu',-\kappa)> \quad , \tag{3.23}$$

and the mean square phonon amplitudes become:

$$<\tilde{Q}(\mu,\kappa)\,Q(\mu',\kappa)> = \delta_{\mu\mu'}\,\frac{1}{2\bar{\omega}(\mu,\kappa)}\,\coth\left\{\frac{1}{2}\,\beta\bar{\omega}(\mu,\kappa)\right\} \quad . \tag{3.24}$$

From these two equations the self-consistency condition for the renormalized mode frequencies $\bar{\omega}(\lambda,q)$ becomes ($\beta = 1/kT$):

$$\bar{\omega}^2(\lambda,q) = \omega_0^2(\lambda,q) + \sum_{\kappa\mu} g^{(4)}_{\lambda\mu}(q,\kappa)\,\frac{1}{2\bar{\omega}(\mu,\kappa)}\,\coth\left\{\frac{1}{2}\,\beta\bar{\omega}(\mu,\kappa)\right\} \quad , \tag{3.25}$$

where $g^{(4)}_{\lambda,\mu}(q,\kappa) = 2\,V^{(4)}_{\lambda\lambda\mu\mu}(q,\kappa)$.

In order to have a structural change it is necessary that at least one of the renormalized mode frequencies becomes imaginary, $\bar{\omega}^2(\lambda,q_0) < 0$, where $q_0$ is the critical wave vector for which $w_0^2(\lambda,q_0)$ has its largest negative value. Simultaneously its negative value cannot be so large (Pytte, 1972) that anharmonic thermal fluctuations could not stabilize the paraelectric phase.

Instead of solving the self-consistency problem very often the quasi-harmonic approximation is used (Silverman and Joseph, 1963,1964; Silverman, 1963,1964; Cowley, 1965; Kwok and Miller, 1966) in which the unknown frequencies in the summation $\bar{w}(\lambda,q)$ are replaced by the known harmonic frequencies:

$$\bar{w}^2(\lambda,q) = w_0^2(\lambda,q) + \sum_{\mu\kappa} g_{\lambda\mu}^{(4)} (q,\kappa) \frac{1}{2w_0(\mu,\kappa)} \coth\left\{\frac{1}{2} \beta w_0(\mu,\kappa)\right\} \quad . \tag{3.26}$$

In the limit of high temperatures, $\beta w_0(\mu,k) << 1$, and for $q=q_0$, the temperature dependence of the renormalized unstable mode becomes

$$\bar{w}^2(\lambda,q_0) = k \sum_{\kappa\mu} g_{\lambda\mu}^{(4)} (q_0,\kappa) \frac{1}{w_0^2(\mu,\kappa)} (T - T_\lambda) \quad , \tag{3.27}$$

where $T_\lambda$ is given by

$$T_\lambda = - \frac{w_0^2(\lambda,q_0)}{k \sum_{\kappa\mu} g_{\lambda\mu}^{(4)} (q_0,\kappa)} w_0^2(\mu,\kappa) \quad . \tag{3.28}$$

This results in a Curie-Weiss-law for the static dielectric constant $\varepsilon_0$. The critical wave vector $q_0$ is equal to zero for a ferroelectric phase transition and for $q_0 \neq 0$ either an antiferroelectric state results or an incommensurate one.

To evaluate the free energy the starting Hamiltonian is transformed to

$$H = \sum_q \frac{1}{2} \left[P_q^2 + \bar{w}^2(q)Q_q^2\right] + E(T) \quad , \tag{3.29}$$

where $P_q$ and $Q_q$ are the momentum and coordinate operators for a normal mode with wavevector q and E(T) is a temperature dependent constant. Assuming a classical system the probability density in phase space can be written as a product of probability densities in momentum $\rho(P_q)$ and configuration space $\rho(Q_q)$.

The equilibrium values of both are then

$$\rho_0(P_q) = \frac{1}{\sqrt{2\pi kT}} \exp(-P_q^2/2kT) \quad , \tag{3.30}$$

and

$$\rho_0(Q_q) = \frac{\bar{w}(q)}{\sqrt{2\pi kT}} \exp[-\bar{w}^2(q)Q_q^2/2kT] \quad . \tag{3.31}$$

$\rho(P_q)$ can be approximated by its equilibrium value $\rho_0(P_q)$. $\rho(Q_q)$ is given by a displaced harmonic oscillator probability function, i.e.

$$\rho(Q_q) = \frac{\bar{w}(q)}{\sqrt{2\pi kT}} \exp[-\bar{w}^2(q)(Q_q - Q_{q,0})^2/2kT] \quad , \tag{3.32}$$

67

where $Q_{q,o}$ is the mean value of the phonon eigenvector in the non-equilibrium state, which becomes the order parameter for the critical wave vector $q = q_o$. The free energy of the system $\phi$ can now easily be evaluated with

$$\langle H \rangle = \sum_q \left[ kT + \frac{1}{2} \bar{\omega}^2(q) \, Q^2_{q,o} \right] \quad , \tag{3.33}$$

and

$$S = \frac{1}{2} \sum_q k \ln kT/\bar{\omega}^2(q) + \text{constant terms}, \tag{3.34}$$

to be

$$\phi^{(2)} = \frac{1}{2} \sum_q \bar{\omega}^2(q) \, Q^2_{q,o} \quad , \tag{3.35}$$

which for $q = q_o$ yields for $\bar{\omega}^2(q_o)$

$$\bar{\omega}^2(q_o) = k(T - T_o) \quad , \tag{3.36}$$

where $T_o$ corresponds to the phase transition temperature.

It has to be pointed out that the basic result of the above treatment, $\bar{\omega}^2(\lambda, q_o) \approx (T - T_c)$ is independent of whether the quasi-harmonic approximation is used or the self-consistent procedure is applied.


## 3.3.2 The Molecular-Field Treatment of a Single Mode Model

In the previous sections only the coupling of the soft phonon with the other non-soft phonons has been taken into account. Another approach to the problem of soft modes can be made by studying the coupling of the soft mode with itself, a treatment which has first been proposed by Thomas (1969,1971) and Thomas and Müller (1972). The starting Hamiltonian describes a system with one degree of freedom per unit cell and a lattice instability which is caused by an optical phonon:

$$H = \sum_{\ell} \left\{ \frac{1}{2M} P^2_\ell + V(Q_\ell) \right\} - \frac{1}{2} \sum_{\ell\ell'} v_{\ell\ell'} Q_\ell Q_{\ell'} \quad . \tag{3.37}$$

As before $Q_\ell$ is a local normal coordinate of the ion displacement in the $\ell'^{th}$ unit cell and $P_\ell$ are the respective momenta. $v_{\ell\ell'}$ describes the interaction between the ionic displacements in different unit cells. $V(Q_\ell)$ is a single particle potential which consists of two terms

$$V(Q_\ell) = V_2 + V_4 = \frac{1}{2} M \omega^2_o Q^2_\ell + \frac{1}{4} \gamma Q_\ell^4 \quad , \tag{3.38}$$

where the first term describes the pure harmonic contribution while the second term adds the anharmonicity. The potential $V(Q_\ell)$ has a single minimum at $Q = 0$ for $\omega^2_o$, $\gamma > 0$.

For strong enough interactions $v_{\ell\ell'} Q_\ell Q_{\ell'}$ the equilibrium position at $Q_\ell = 0$ can be destabilized thus inducing a displacive phase transition. For a discussion

68

of the static properties of the system the density matrix of the many-body system is written as the product of single particle density matrixes

$$\rho = \prod_{\ell} \rho_{\ell} \quad \text{with} \quad \text{Tr}\{\rho\} = \text{Tr}\{\rho_{\ell}\} = 1 \quad . \tag{3.39}$$

The equilibrium properties of the system are determined by the minimization of the free energy with respect to $\rho_{\ell}$:

$$\frac{dF}{d\rho_{\ell}} = \text{Tr}\left\{ \frac{1}{2M} P_{\ell}^2 + V(Q_{\ell}) - \sum_{\ell} v_{\ell\ell'} Q_{\ell} <Q_{\ell'}> + kT \ln \rho_{\ell} + kT \right\} = 0 \quad , \tag{3.40}$$

where $<Q_{\ell}> = Q_0$ for all $\ell$ is the thermal average of the soft mode coordinate. From this the single particle molecular field Hamiltonian is given by

$$H_{MFA} = - \frac{1}{2M} P_{\ell}^2 + V(Q_{\ell}) - H_{\ell} Q_{\ell} \quad \text{with} \quad H_{\ell} = \sum_{\ell'} v_{\ell\ell'} <Q_{\ell'}> \quad . \tag{3.41}$$

Even in this simplified version it is not possible to find analytically solutions of the above outlined problem. Yet if one uses instead of the exact probability densities $\rho_{\ell}(P_{\ell}, Q_{\ell})$ displaced harmonic oscillator type trial probability densities one can find the following solutions: $<Q_{\ell}> = Q_0 = 0$ which obviously applies to the paraelectric phase. In the low temperature ordered phase one gets

$$<Q>_{1,2}^2 = \frac{1}{4\gamma} \left[ v_0 - 2M\omega_0^2 \pm (3v_0 - 2M\omega_0^2)^2 - 24\gamma kT \right]^{\frac{1}{2}} \quad , \tag{3.42}$$

where $v_0 = \sum_{\ell'} v_{\ell\ell'}$ is independent of $\ell$.

For the discussion of the dynamic properties a small internal time dependent field $E_{\ell} e^{-i\omega t}$ is added to the original Hamiltonian so that H becomes

$$H(t) = \sum_{\ell} \left\{ \frac{1}{2M} P_{\ell}^2 + V(Q_{\ell}) \right\} - \frac{1}{2} \sum_{\ell\ell'} v_{\ell\ell'} Q_{\ell} Q_{\ell'} - \sum_{\ell} E_{\ell} Q_{\ell} e^{-i\omega t} \quad . \tag{3.43}$$

Again it is assumed that the exact probability density can be approximated by a displaced harmonic oscillator density

$$\rho_{\ell}(Q_{\ell}, t) = \frac{1}{\sqrt{2\pi\sigma_{\ell}}} \exp\left\{ -(Q_{\ell} - <Q_{\ell}>_t)^2 / 2\sigma_{\ell} \right\} \quad , \tag{3.44}$$

$$\rho_{\ell} = < (Q_{\ell} - <Q_{\ell}>)^2 > \quad . \tag{3.45}$$

Now the equations of motion for the time dependent expectation values $<Q_{\ell}>$ become:

$$M \frac{d^2}{dt^2} <Q_{\ell}>_t = - M \omega_0^2 <Q_{\ell}>_t - \gamma<Q_{\ell}^3>_t - 3 \gamma\sigma<Q_{\ell}>_t$$

$$+ \sum_{\ell\ell'} v_{\ell\ell'} <Q_{\ell'}>_t + E_{\ell} e^{-i\omega t} \quad . \tag{3.46}$$

69

Assuming that the normal mode coordinates oscillate around their MFA average values
with the frequency of the external field

$$<Q_\ell> = <Q> + \delta <Q_\ell> e^{-i\omega t} \quad , \tag{3.47}$$

the effective normal mode frequencies $\bar{\omega}(q)$ are given by:

$$M \bar{\omega}^2(q) = M \omega_0^2 + 3\gamma\sigma + 3\gamma<Q>^2 - \sum_{\ell'} V_{\ell\ell'} \exp[iq(R_\ell - R_{\ell'})] \quad . \tag{3.48}$$

The destabilization of the system thus enters via the harmonic intercell interac-
tions $\sum_{\ell'} V_{\ell\ell'} \exp[iq(R_\ell-R_{\ell'})]$. Again one ends up with a self-consistent equation for
the renormalized temperature dependent normal modes $\bar{\omega}(q)$ which has the same form
as the self-consistent equation for the case that $\bar{\omega}(\lambda,q)$ couples to all other modes
except the soft ones. To overcome the difficulties included in the self-consistent
procedure Gillis and Koehler (1972) proposed to replace $\bar{\omega}(q)$ by an effective single
particle frequency $\omega_s$ such that

$$<(Q - <Q> )^2> = \rho \approx \frac{kT}{M \omega_s^2} \quad , \tag{3.49}$$

which leads for $T > T_0$, where $<Q> = 0$, to

$$M \bar{\omega}^2(q) = M \omega_0^2 + 3\gamma\sigma - \sum_{\ell\ell'} V_{\ell\ell'} \exp[iq(R_\ell - R_{\ell'})] \quad . \tag{3.50}$$

The transition will take place, if the Fourier transform of the harmonic intercell
interaction $v_q = \sum_{\ell'} V_{\ell\ell'} \exp[iq(R_\ell-R_\ell)] = v_{q_0} = M\omega_0^2 + 3\gamma\sigma$ which yields for the transition temperature $T_0$

$$k T_0 = \frac{1}{3\gamma} v_{q_0} (v_{q_0} - M\omega_0^2) \quad . \tag{3.51}$$

The temperature dependence of the soft mode can now be expressed by

$$M \bar{\omega}^2 (q) = \frac{3\gamma k}{2 v_{q_0} - M \omega_0^2} (T - T_0) + (v_{q_0} - v_q) \quad \text{for} \quad T > T_0 \quad , \tag{3.52a}$$

and

$$M \bar{\omega}^2 (q) = \frac{1}{2} \left\{ M \omega_0^2 - 2 v_q + \left[ (M \omega_0^2)^2 + 12\gamma kT \right]^{\frac{1}{2}} \right\} \quad \text{for} \quad T < T_0 \quad . \tag{3.52b}$$

Within the above outlined models it is possible to reproduce qualitatively the experimental data on the temperature dependence of the soft-mode. Yet quantitative
agreement could not be achieved till now as either anharmonic effects of higher
order have to be taken into account or the microscopic origin of the lattice instability has to be reconsidered.

70

The vibronic model has first been established by Kristoffel, Konsin (1967,1968, 1969,1971,1972), Bersuker and Vekhter (1967,1969a,1969b) to explain the occurrence of ferroelectricity in crystals with extremely small energy band gaps, a situation which is found for instance in SbSI and the IV-VI-semiconductors. Consider again the effective Hamiltonian of the anharmonic lattice model for ionic motion. It consists mainly of three parts:

$$H_{eff}(ion) = \sum_i \frac{p_i^2}{2m_i} + U(R_i,R_j) + E(R_i,R_j,...) \quad , \tag{3.53}$$

where the first two terms represent the kinetic and potential energies of the ionic cores while $E(R_i,R_j...)$ represents the interaction energy of the valence electrons and the ions, the vibronic energy. In the anharmonic lattice model it is assumed that $E(R_i,R_j...)$ is independent of temperature which means that the energy separation between valence and conduction band is large enough to neglect thermal excitations between the two bands. For narrow-gap materials this situation obviously does not apply any more. To describe ferroelectric phase transitions in these materials properly, interactions from the valence to the conduction band have to be taken into account too.

The model which is considered in the vibronic theory, starts from two non-degenerate electronic bands with energies $\varepsilon_1(q)$ and $\varepsilon_2(q)$. At $T = 0$ the lower band, the valence band, $\varepsilon_1$ is fully occupied while the upper conduction band, $\varepsilon_2$, is empty. Vibronic interactions take place if these bands are mixed by suitable lattice vibrations. Since the uniform lattice mode distortion $y_q$, which in the ferroelectric low-symmetry phase induces the spontaneous polarization, is of odd symmetry it follows that intraband transitions can only take place in even order. Thus the eigenvalue problem becomes:

$$\begin{vmatrix} \varepsilon_1(q) - E & \frac{1}{\sqrt{N_o}} \sum_q V(q)y_q \\ \frac{1}{\sqrt{N_o}} \sum_q V(q)y_q & \varepsilon_2(q) - E \end{vmatrix} \begin{vmatrix} A \\ B \end{vmatrix} = 0, \tag{3.54}$$

where $V(q)$ is the interband electron-phonon interaction term and $N_o$ the number of unit cells in the crystal. Accordingly the perturbed electronic band energies are:

$$E_{\pm} = \frac{1}{2}\left[\varepsilon_1(q) + \varepsilon_2(q)\right] \pm \left\{\frac{1}{4}\left[\varepsilon_1(q) - \varepsilon_2(q)\right]^2 + \frac{1}{N_o}\sum_q V^2(q)y_q^2\right\}^{\frac{1}{2}} \quad . \tag{3.55}$$

The Helmholtz free energy of the system becomes:

$$F(T,y_o) = - kT \sum_q \ln \left\{ 2 + 2 \cosh \frac{1}{kT} \left[ \frac{[\epsilon_1(q)-\epsilon_2(q)]^2}{4} + \frac{1}{N_o} v^2(0)y_o^2 \right]^{\frac{1}{2}} \right\}$$
$$+ \frac{1}{2} M\omega_o^2 y_o^2 \quad , \tag{3.56}$$

with M being the reduced mass corresponding to the active vibrations with the initial value of the frequency $\omega_o$.

A phase transition to a low symmetry ferroelectric phase will take place if

$$\frac{\partial F(T,y_o)}{\partial y_o} = 0 \tag{3.57}$$

has a real solution $y_o \neq 0$. Direct differentiation of Eq. (3.56) yields

$$y_o^2 = \frac{1}{N_o} \frac{v^2}{\omega_o 4} \tanh^2 \left\{ \frac{1}{4kT} \left[ (\epsilon_2 - \epsilon_1)^2 + 4 \frac{v^2}{N_o} y_o^2 \right]^{\frac{1}{2}} \right\} - \frac{(\epsilon_2-\epsilon_1)^2 N_o}{4v^2} . \tag{3.58}$$

For high temperature, i.e. $(\epsilon_2 - \epsilon_1)/kT \to 0$.

$$y_o^2 \to - \frac{(\epsilon_2 - \epsilon_1)^2 N_o}{4v^2} \tag{3.59}$$

so that a lattice distortion cannot take place. For very low temperatures $y_o^2$
becomes

$$y_o^2 = \frac{1}{N_o} \frac{v^2}{\omega_o 4} - \frac{(\epsilon_2 - \epsilon_1)^2 N_o}{4v^2} \quad , \tag{3.60}$$

which has real solutions if $2 \frac{1}{N_o} v^2 > \omega_o^2(\epsilon_2-\epsilon_1)$. The transition temperature
$T_c$ is determined via the condition $y_o^2 = 0$, i.e.

$$kT_c = \frac{\epsilon_2 - \epsilon_1}{4} \left[ \text{artanh} \frac{\omega_o^2(\epsilon_2 - \epsilon_1)N_o}{2v^2} \right]^{-1} . \tag{3.61}$$

From this equation the "softness" of the system under consideration can be determined. The smaller the gap and the stronger the electron-phonon interaction the easier a transition to a distorted structure can take place. Furthermore it has to be pointed out that in this model the essential anharmonicity of the system is caused by the vibronic interaction.

Via the temperature dependent vibronic interaction term the soft phonon is renormalized and becomes temperature dependent too.

Yet the calculations of the temperature dependence of the soft ferroelectric mode have shown that the vibronic interaction term is not sufficient to induce a soft mode behaviour in the IV-VI compounds. To overcome this difficulty anharmonic lattice interactions have to be included in the Hamiltonian, i.e. via only phonon-

phonon anharmonicity the strong temperature dependence of $\omega_f^2$ can be achieved. Nevertheless it is not possible to reproduce the experimental data of the soft mode frequency by means of the vibronic model including lattice anharmonicity. Obviously, as will be shown in the next paragraph, not phonon-phonon anharmonicity has to be considered but electron-phonon anharmonicity.

In the IV-VI semiconductors free carriers, extrinsic ones as well as intrinsic, play an important role. By varying the free carrier concentration the phase transition temperature can be shifted and even the phase transition itself can be completely suppressed. This important effect has been taken into account within the vibronic model.

Assuming that the system under consideration has only extrinsic impurities these induce additional terms in the energies which are perturbed by vibronic interactions. Each additional electron from a donor ion adds a potential energy term $E_+$, and each valence band hole reduces the energy by $E_-$. Thus for a small number $N_e$ of carriers and $N_h$ of holes compared to the number $N$ of energy states the electronic potential becomes:

$$E(y_o) = \frac{1}{2}(N_e + N_h - N)\left[(\varepsilon_1 - \varepsilon_2)^2 + 4\frac{1}{N_o}v^2 y_o^2\right]^{\frac{1}{2}} . \qquad (3.62)$$

For intrinsic impurities, which are thermally created electrons and holes, $N_e$ and $N_h$ are the same and the energy can be written as

$$E(y_o) = -\frac{1}{2}N\left\{\left[\varepsilon_2(q) - {}_2(q)\right]^2 + 4\frac{1}{N_o}v^2 y_o^2\right\}^{\frac{1}{2}} , \qquad (3.63)$$

which yields for the contribution from the carriers alone:

$$E_c = +\frac{1}{2}(N_e + N_h)\left\{\left[\varepsilon_1(q) - \varepsilon_2(q)\right]^2 + 4\frac{1}{N_o}v^2 y_o^2\right\}^{\frac{1}{2}} . \qquad (3.64a)$$

For small displacements $y_o$, $E_c$ can be expanded

$$E_c = \frac{1}{2}(N_e + N_h)\left\{\left[\varepsilon_1(q) - \varepsilon_2(q)\right]^2 + \frac{\cdot\ 2v^2 y_o^2}{N_o[\varepsilon_1(q) - \varepsilon_2(q)]}\right.$$

$$\left. - \frac{2v^4 y_o^4}{N_o^2[\varepsilon_1(q) - \varepsilon_2(q)]^3} + \ldots\right\} , \qquad (3.64b)$$

which leads to the obvious conclusion that the additional carriers stiffen the lattice and stabilize the paraelectric phase. The transition temperature reduces to lower temperatures in the presence of carriers. Although the vibronic model is qualitatively in agreement with the experimental data, one is not able to achieve even a rough quantitative correspondence to the available data.

More than 98% of all crystals which exhibit a soft mode behaviour, that can be related to a ferro-or antiferroelectric phase transition, contain either oxygen or
other chalcogenide ions. This fact suggests a central role of chalcogenide ions in
ferroelectric phase transitions. For perovskites (Migoni et al., 1976), SbSI (Balkanski et al., 1980) and $K_2SeO_4$ (Bilz et al., 1982) it has been shown that the quartic highly anisotropic oxygen ion polarizability governs their soft mode behaviour.
Within a simple diatomic linear chain model (Bilz et al., 1980) with non-linear polarizability at the chalcogenide ion lattice site it was possible to describe the
dispersion of the lowest transverse acoustic and the transverse ferroelectric optic mode as well as the temperature dependence of $\omega_f$ at q = 0. For the IV-VI  semiconducting compounds it was possible to use the same model to describe the temperature dependence and the dispersion of $\omega_f(q)$. In this case the $S^{2-}$, $Se^{2-}$ and $Te^{2-}$
ion polarizabilities represent the driving mechanism of the ferroelectric phase
transition in these compounds (Bussmann-Holder et al., 1980,1981a).

   The oxygen ion $O^{2-}$ and its homologues are unstable as free ions. In a crystal
they are stabilized via the interaction with the surrounding ions, that means the
Madelung potential. The evaluation of phenomenological data on the oxygen ion polarizability  $\alpha$ has shown (Tessmann et al., 1953) that in simple cubic oxides
(MgO, etc.) $\alpha$ is proportional to the volume V of the ion. In tetrahedral oxides,
as e.g. ZnO, the covalent bonding leads to a $V^2$ dependence of $\alpha$ . A further enhancement of the volume dependence of the polarizability may be due to anisotropy as
has been observed in the spinels where $\alpha$ is proportional to $V^4$ (Kirsch et al.,
1974).

   To calculate quantum-mechanically the oxygen ion polarizability, the stabiliz-
ing effect of the crystalline Coulomb potential can be simulated by a homogeneously
(+ 2e) charged sphere, the Watson sphere (Watson, 1958), the potential of which is
equal to the Madelung potential of the crystal. The variation of the radius of the
sphere, the Watson radius $R_w$, corresponds to a variation of the lattice constant.

   Within the isotropic Watson model it is possible to compute the wave functions
and charge densities of the respective ion and furthermore by means of the forma-
lism of Thorhallson et al. (1968) the polarizability (Bussmann et al., 1980). In
Fig. 3.1 the polarizabilities to $F^-$ and $O^{2-}$ are shown as a function of the Watson
radius. While  $\alpha$ of $F^-$ converges to its free ion value for $R_w \to \infty$, $\alpha(O^{2-})$ diverges.
In the physically relevant range of $R_w$, $(O^{2-})$ varies approximately as $R_w^3$, which
means that the Watson model reproduces closely the variation of $\alpha$  in simple oxides. The enhancement of the volume dependence of  $\alpha(O^{2-})$, which has been observed
in $ABO_3$ and spinels (Kirsch, 1974) due to anisotropy and covalency, can be taken
into account by assuming an ellipsoidal charge distribution. This effect can be
studied in the Watson model by taking a weighted average over Watson spheres with

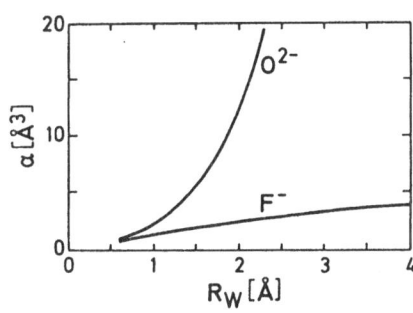

Fig. 3.1 Polarization of $0^{2-}$ and $F^-$ ions as a function of the Watson radius $R_W$

different radii $R_1$ and $R_2$, which leads to a further volume dependence of $\alpha$, additional to the volume dependence of the isotropic case.

The strong volume dependence of $\alpha(0^{2-})$ plays the essential role in ferroelectric systems and triggers the phase transition. The "pathological" behaviour of $0^{2-}$ and its homologues has been taken into account in a linear diatomic shell model (Bilz et al., 1980). While the cation is assumed to be rigid or only weakly polarizable $(g_2^{(2)})$, the anionic core-shell-coupling consists of a strongly attractive harmonic electron ion coupling constant $g_2^{(1)}$ and a non-linear repulsive term $g_4$ (Fig. 3.2).

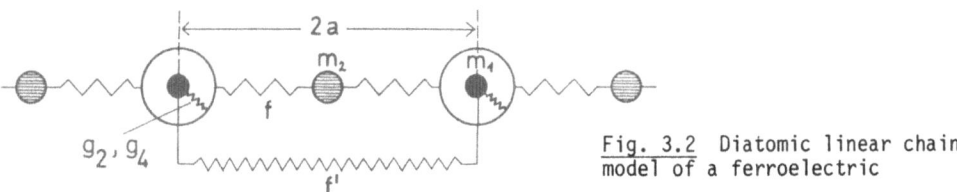

Fig. 3.2 Diatomic linear chain model of a ferroelectric

The instability of the ferroelectric soft mode is attributed to the attractive Coulomb forces (modeled by $g_2$). Its stabilization in the paraelectric and the ferroelectric regime is achieved via the on-site fourth-order core shell coupling $g_4$ and the second nearest neighbour core-core couplings $f'$ and $f''$.

The model Hamiltonian reads

$$H = T + V \tag{3.65}$$

where

$$T = \frac{1}{2} \sum_n (m_1 \dot{u}_{1n}^2 + m_2 \dot{u}_{2n}^2 + m_{e1} \dot{v}_{1n}^2 + m_{e2} \dot{v}_{2n}^2) \quad , \tag{3.66}$$

and

$$V = \frac{1}{2} \sum_n \left[ f'(u_{1n} - u_{1n-1})^2 + f(v_{2n} - v_{1n})^2 + f''(u_{2n} - u_{2n-1})^2 \right.$$
$$+ f(v_{2n} - v_{1n+1})^2 + g_2^{(1)}(v_{1n} - u_{1n})^2 + g_2^{(2)}(v_{2n} - u_{2n})^2$$
$$\left. + \frac{1}{2} g_4 (v_{1n} - u_{1n})^4 \right] \quad . \tag{3.67}$$

75

$m_1$, $m_{e1}$, $m_2$, $m_{e2}$ are the core and shell masses of atoms 1 and 2 and $u_{1n}, v_{1n}, u_{2n}, v_{2n}$ their respective displacements in the $n^{th}$ cell.

By defining a new difference coordinate $w_{in} = v_{in} - u_{in}$ ($i = 1,2$) V can be re-written in the following way:

$$V = \frac{1}{2} \sum_n \left[ f'(u_{1n} - u_{1n-1})^2 + f''(u_{2n} - u_{2n-1})^2 \right.$$
$$+ f(u_{2n} - u_{1n} + w_{2n} - w_{1n})^2 + f(u_{2n} - u_{1n+1} + w_{2n} - w_{1n+1})^2$$
$$\left. + g_2^{(2)} w_{2n}^2 + g_2^{(1)} w_{1n}^2 + \frac{1}{2} g_4 w_{1n}^4 \right] . \tag{3.68}$$

The equations of motion contain the first derivatives of the potential with respect to the displacements, where for the shells the adiabatic condition is used:

$$m_1 \ddot{u}_{1n} = f'(u_{1n+1} + u_{1n-1} - 2 u_{1n}) + g_2^{(1)} w_{1n} + g_4 w_{1n}^3 , \tag{3.69}$$

$$m_2 \ddot{u}_{2n} = f''(u_{2n+1} + u_{2n-1} - 2 u_{2n}) + g_2^{(2)} w_{2n} , \tag{3.70}$$

$$0 = - g_2^{(1)} w_{1n} - g_4 w_{1n}^3 - 2f w_{1n} + f(u_{2n} + u_{2n-1}$$
$$- 2 u_{1n} + w_{2n} + w_{2n-1}) , \tag{3.71}$$

$$0 = - g_2^{(2)} w_{2n} - 2f w_{2n} + f (u_{1n} + u_{1n-1} - 2 u_{2n} + w_{1n} + w_{1n-1}) . \tag{3.72}$$

This model has some features in common with the model ferroelectrics discussed by Pytte (1972), Gillis and Koehler (1972), and Eisenriegler (1974) since it is also a double-well problem. It differs, however, from them in important details such as a local quartic electron-ion coupling which accounts for the high polarizability of the chalcogenide ions. Furthermore the model exhibits exact non-linear solutions by means of which it is possible to explain phonon anomalies which have been observed in several different ferroelectric systems.

For the description of the temperature dependence of the soft-mode and related quantities it is necessary to study the approximate solutions in the selfconsistent phonon approximation (SPA).

The SPA corresponds to a linearization of the cubic term, which enters via $g_4$ in the equations of motion, i.e.

$$g_4 w_{1n}^3 \overset{(SPA)}{=} 3 g_4 w_{1n} <w_{1n}^2>_T = g_T w_{1n} , \tag{3.73}$$

where

$$< w_{1n}^2 >_T = \sum_{\vec{q}j} \frac{\hbar}{2Nm_1 \omega_{qj}} w_1^2(\vec{q}j) \coth \frac{\hbar \omega_{qj}}{2kT} . \tag{3.74}$$

By means of this approximations Eqs. (3.69)-(3.72) now read:

$$m_1 \ddot{u}_{1n} = f' \, (u_{1n+1} + u_{1n-1} - 2 \, u_{1n}) + g(T) \, w_{1n} \quad , \tag{3.75}$$

$$0 = - \, (g(T) + 2f) \, w_{1n} + f \, (u_{2n} + u_{2n-1} - 2 \, u_{1n} + w_{2n} + w_{2n-1}) \quad , \tag{3.76}$$

where

$$g(T) = g_2^{(1)} + 3g_4 \, <w_{1n}^2> \, T \quad . \tag{3.77}$$

In the paraelectric regime $(T > T_c)$ $g(T)$ is positive, while in the ferroelectric regime $g(T)$ is replaced by $- \, 2g(T)$, that means it behaves like the parameter a in Landau theory [see Eq.(1.1)].

Within the SPA stable solutions are obtained from the equations of motion which lead to the following equations in q-space by using standard Fourier transform technique:

$$\omega^2 \begin{vmatrix} m_1 & 0 \\ 0 & m_2 \end{vmatrix} \begin{vmatrix} u_1 \\ u_2 \end{vmatrix} = D(q) \begin{vmatrix} u_1 \\ u_2 \end{vmatrix} \quad , \tag{3.78}$$

where the dynamical matrix is given by

$$D(q) = 2\tilde{f} \begin{vmatrix} 1 + A_1 \sin^2 qa & - \cos qa \\ - \cos qa & 1 + A_2 \sin^2 q_a \end{vmatrix} \quad , \tag{3.79}$$

with

$$\tilde{f} = \frac{f \, g(T) g_2^{(2)}}{[2f + g(T)](2f + g_2^{(2)}) - 4f^2 \cos^2 qa} \quad , \tag{3.80}$$

$$A_1 = \frac{4f'}{2\tilde{f}} + \frac{2f}{g_2^{(2)}} \quad , \quad A_2 = \frac{4f''}{2f} + \frac{2f}{g(T)} \quad . \tag{3.81}$$

The dispersion relation can be written explicitly as:

$$\omega_{\pm}^2 = \tilde{f} \left[ \frac{1}{m_1} \, (1 + A_1 \sin^2 qa) + \frac{1}{m_2} \, (1 + A_2 \sin^2 qa) \right]$$

$$\pm \left\{ \left[ \frac{1}{m_1} \, (1 + A_1 \sin^2 qa) - \frac{1}{m_2} \, (1 + A_2 \sin^2 qa) \right]^2 + \frac{4 \cos^2 qa}{m_1 \, m_2} \right\}^{\frac{1}{2}} \quad , \tag{3.82}$$

where in the limit $q \rightarrow 0$

$$\omega_-^2 \cong \left[ \frac{1}{m_1} \, (2f + 4f') + \frac{1}{m_2} \, (2f + 4f'') \right] \sin^2 qa \quad , \tag{3.83}$$

and the ferroelectric mode

$$\omega_+^2 \equiv \omega_f^2 = \omega_0^2 \, \frac{1}{1 + 2f \left[ \frac{1}{g(T)} + \frac{1}{g_2^{(2)}} \right]} \quad , \tag{3.84}$$

77

$$\omega_0^2 = \frac{2f}{\mu} \quad , \tag{3.85}$$

with $\mu$ being the reduced mass and $\omega_0^2$ the rigid-ion limit of the ferroelectric mode for $g(T)$ and $g_2^{(2)} \to \infty$.

The zone boundary frequencies are given by:

$$\omega_+^2 \left(\frac{\pi}{2a}\right) = \frac{2}{m_2} \left(2f'' + \frac{fh}{2f+h}\right) \quad , \tag{3.86}$$

$$\omega_-^2 \left(\frac{\pi}{2a}\right) = \frac{2}{m_1} \left[2f' + \frac{fg(T)}{2f+g(T)}\right] \quad , \tag{3.87}$$

where $h \equiv g_2^{(2)}$. At the phase transition $g(T) = 0$ and $\omega_f^2(q=0) = 0$. Near $T_c$ the ferroelectric optic branch becomes a pseudo-acoustic mode since the system splits now into two independent chains. Thus for small q the new dispersion relations for the two acoustic modes read:

$$\omega_1^2 \cong \frac{4f'}{m_1} \sin^2 qa \quad , \tag{3.88}$$

$$\omega_2^2 \cong \frac{1}{m_2} (2f + 4f'') \sin^2 qa \quad . \tag{3.89}$$

For finite $g(T)$, i.e. $T > T_c$, obviously $\omega_+^2$ is temperature dependent for small wave vectors, while $\omega_-^2$ becomes temperature dependent for large q. This can be taken into account by transforming the dynamical matrix such that one branch contains approximately all the temperature dependence while the other is nearly T-independent. The new resulting interpolating modes $\omega_f^2(q,T)$ and $\omega_A^2(q)$ tend, for $q \to 0$, to the original modes $\omega_+^2$ and $\omega_-^2$, respectively, while they are interchanged at the zone boundary. By means of the above proposed transformation $\omega_+^2$ and $\omega_-^2$ now become

$$\omega_f^2 = \frac{2\tilde{f}}{\mu} + \left(\frac{2f}{h}\frac{2\tilde{f}}{m_1} + \frac{4f'}{m_1} - \frac{2\tilde{f}}{m_2}\right) \sin^2 qa \quad , \tag{3.90}$$

$$\omega_A^2 = \frac{2\tilde{f}}{m_2} \left[1 + \frac{2f}{g(T)} + \frac{4f''}{2\tilde{f}}\right] \sin^2 qa \quad . \tag{3.91}$$

Within this transformation the off-diagonal elements of the new dynamical matrix have been neglected. This can be easily justified as their contribution to the dynamical matrix is equal to zero at the zone center and the zone boundary, while it is of the order of $g(T)^2$ in between. When $g(T) \to 0$, i.e. when one approaches the phase transition at $T_c$, the new description becomes exact.

In all ferroelectric compounds $\omega_f^2$ is a function of temperature and of the phase transition point, the Curie temperature $T_c$. Generally this dependence is expressed by

$$\omega_f^2 = \alpha (T - T_c)^\gamma \quad , \tag{3.92}$$

where $\alpha$, $T_c$ and $\gamma$ have to be determined via the internal model parameters. Within the polarizability model the temperature dependence of the ferroelectric soft mode enters via the self-consistent thermal average over the relative core-shell displacement squared at the anionic lattice site, $<w_{1n}^2>_T$. Since $<w_{1n}^2>_T$ is a function of $\omega_f^2$ and $T$, a single self-consistent equation which gives the implicit relation between $\omega_f^2$ and $T$, results:

$$g_2^{(1)} \left( 1 - \frac{\omega_f^2}{\omega_0^2} \right) + 3 g_2^{(1)} \left( 1 - \frac{\omega_f^2}{\omega_0^2} \right)^3 I_F(\omega_f, T) = \mu \omega_f^2 \quad , \tag{3.93}$$

where $I_F$ is an integral over the three-dimensional Brillouin zone containing all the dynamical information. To study the analytical behaviour of $\omega_f$ for different temperature regimes it is sufficient to consider only the ferroelectric branch with an isotropic dispersion and a Debye-like behaviour for the calculation of $<w_{1n}^2>_T$.

$$I_F(\omega_f, T) \cong \frac{3kT}{2\mu\omega_D^3} \int_0^{\omega_D} d\omega \frac{\omega^2}{(\omega_f^2 + \omega^2)^{1/2}} \coth \frac{\hbar(\omega_f^2 + \omega^2)^{1/2}}{2kT} \quad , \tag{3.94}$$

where $\omega_D^2 = \pi^2 f'/m_1$. The analytical behaviour of $I_F$ determines the T-dependence of $\omega_f^2$ in different regimes.

In order to describe the temperature dependence of $\omega_f^2$ the temperature interval can be roughly divided into four segments.

For temperatures far away from the actual phase transition point $\omega_f^2$ tends to its rigid-ion value and saturates according to the asymptotic expression:

$$\omega_f^2 \approx \omega_0^2 - \text{const.} \ T^{-1/3} \quad . \tag{3.95}$$

Experimentally the effect of saturation has been observed in some perovskites (Müller et al., 1979) and SbSI (Balkanski et al., 1980).

The well-known mean-field regime, where the critical exponent $\gamma$ is equal to one, appears for temperatures $T \gg \omega_f, \omega_D$ and for $\omega_f^2 \ll \omega_0^2$. The evaluation of the integral yields for this region:

$$\omega_f^2 \approx \frac{|g_2^{(1)}|}{3 \mu T_c} (T - T_c) \quad , \qquad \text{where } T_c = - \mu \omega_D^2 g_2^{(1)} / 9 \hbar g_4 . \tag{3.96}$$

The mean-field regime has been observed in all known ferroelectric compounds and until the discovery of deviations from the ($\gamma = 1$) - behaviour of the soft mode, it was believed that $\gamma = 1$ is a unique exponent for all temperatures.

An enhancement of the critical exponent $\gamma$ to values of 1,4 - 1,6 occurs for temperatures close to $T_c$ (Rytz et al., 1980). As in this regime g(T) is small compared to f, $\omega_f^2$ at q = 0 is proportional to g(T). Thus g(T) becomes:

$$g(T) \approx \frac{G}{2}\left(\coth\frac{T_f}{2T} - \frac{2T_c}{T_f}\right) \quad , \tag{3.97}$$

where $T_f$ is the temperature equivalent to $\omega_f$ at the zone boundary and $G = |g_2^{(1)}|/T_c$. This yields directly Barrett's formula (Barrett, 1952) for the dielectric constant $\varepsilon_F$, with empirical parameters A and B,

$$\frac{B}{\varepsilon_F - A} = \frac{2f\ T_f\varepsilon_0/G}{\varepsilon_F - \varepsilon_0} = \frac{T_f}{2}\coth\left(\frac{T_f}{2T}\right) - T_c \quad . \tag{3.98}$$

For $T_c \to 0$ and $T < T_f$, (the quantum limit) renormalization group theory has shown (Schneider et al., 1976) that, simultaneously with an enhancement of the system's dimensionality from d = 3 to d = 4, $\gamma$ is enhanced from 1 to 2. The SPA predicts for this regime

$$\omega_f^2 \approx \left(\frac{T}{T_f}\right)^2 \quad , \tag{3.99}$$

where logarithmic corrections have been neglected.

A renormalization group study of quantum ferroelectrics by Schmeltzer (1983) confirms the result of the SPA for $T_c = 0$. In this study the temperature dependence of the soft mode for $T_c = 0$ is obtained as:

$$\omega_f^2 \approx T^2(\lg T^2)^{-\gamma} \quad , \quad \gamma = 1/3 \quad , \tag{3.100}$$

where the exponent $\gamma$ depends on the model under consideration. It is shown that the self-consistent treatment corresponds directly to renormalization group theory for a large number of components, which explains the considerable success of the SPA, outlined above.

Fig. 3.3 (Bussmann-Holder et al., 1981b) gives a summary of the different temperature regimes which result out of the polarizability model in the SPA including the respective critical exponents for the different regimes. Furthermore it has to be mentioned that the derivation of the effective potential of the model with respect to $w_{1n}^2$ yields for g(T) in the ferroelectric phase a Landau behaviour, i.e.

$$g(T)_{FE} = -2g(T)_{PE} \quad . \tag{3.101}$$

In conclusion, it has to be assumed that the order parameter of the polarizability model is given by $\langle w_{1n}\rangle_T$, the relative core-shell displacement at the anion lattice site, which is zero for $T = T_c$ and 0 for $T < T_c$.

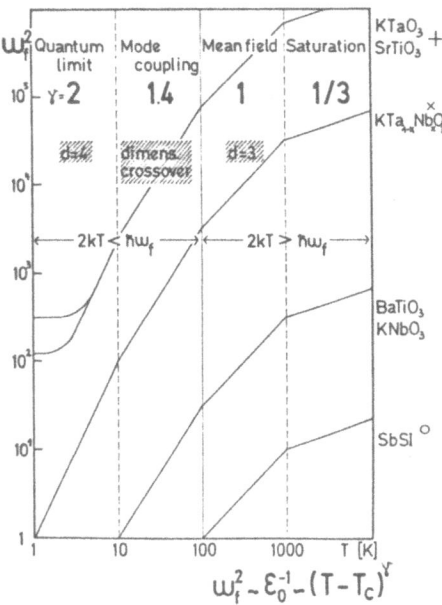

Fig. 3.3 Temperature regimes which result from the model (+ Müller and Burkhard, 1979; x Rytz et al., 1980; o Balkanski et al., 1980)

## 3.5.1 Results

The polarizability model provides the possibility to calculate the dispersion curves for the transverse acoustic and the ferroelectric transverse optic mode and (via the SPA) the temperature dependence of $\omega_f^2(q = 0)$.

The model parameters $g,h,f,f',f''$ have been determined by fitting the theoretical data to the experimental curves. The masses of the diatomic chain were chosen as $m_1$ being the mass of the highly polarizable anion with non-linear core-shell coupling, while $m_2$ was determined by the mass of the weakly polarizable cation. The results of the SPA have been compared to experimental data for PbS, PbSe, PbTe and SnTe.

All the above mentioned compounds exhibit the cubic NaCl-structure in the para-electric phase. As has been shown for perovskites (Bilz et al., 1980), the layer-type model works well for the so called "diagonally cubic" direction, which means a projection of the $ABO_3$ (analogous to the CsCl-type) structure in the (100) di-rection. In the IV-VI compounds the equivalent direction is (111). Therefore the TO and TA mode have been calculated in the (111)-direction. Out of the above men-tioned crystals, only SnTe undergoes a transition to a ferroelectric state with a finite transition temperature of 98 K. The lead chalcogenides belong to the inci-pient ferroelectrics which exhibit a soft-mode behaviour but their transition tem-perature is either close to or even below 0 K. Fig. 3.4 shows the experimental dispersion curves for PbS, PbSe, PbTe and SnTe, compared to the calculated curves. Starting with the parameter set of PbS, the substitution of sulfur by another chalcogenide ion mass affects mainly the TO mode, whose zone boundary and zone

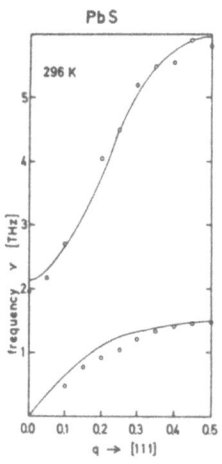

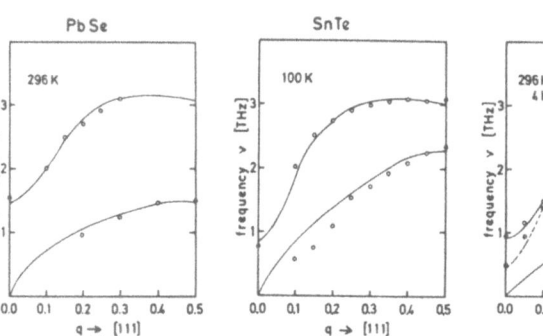

Fig. 3.4 Comparison of the calculated dispersion curves (lines) with the experimental data (dots) for PbS, PbSe, PbTe and SnTe

center frequencies are lowered to smaller values. A slight change of the model parameters leads to the shown theoretical curves. It has to be pointed out that a trend in the increase or decrease of the parameters can be observed with increasing cationic mass. The model parameters are listed in Table 3.1 for the four crystals. For all crystals the second-nearest neighbour core-core coupling constants f' and f" are of the same order of magnitude and the cationic force constant h, being very small, remains constant. The main changes in parameters occur for the next nearest neighbour coupling f, which increases with increasing cationic mass, indicating that the repulsive forces between neighbouring atoms become stronger - a consequence of the increasing overlap between anion and cation. A further remarkable change occurs in g(T), the anionic polarizability, which decreases with increasing difference in masses. As g(T) contains the temperature dependence of the soft mode the decrease in g(T) when going from PbS to SnTe indicates the "softness" of the system, which has also been verified experimentally. PbS [largest g(T)] exhibits only a very small temperature dependence for the q = 0 TO-mode, while SnTe [smallest g(T)] becomes really ferroelectric (Pawley et al., 1966).

Table 3.1 Polarizability model. Model parameters for the lead salts and SnTe. In units of $10^4$ dyn/cm.

|      | PbS   | PbSe  | PbTe  | SnTe  |
|------|-------|-------|-------|-------|
| f    | 4.5   | 12.0  | 13.0  | 18.5  |
| f'   | 1.65  | 0.55  | 0.75  | 0.98  |
| f"   | -1.0  | -2.0  | -1.8  | -2.2  |
| g(T) | 0.95  | 0.85  | 0.45  | 0.3   |
| h    | 28.0  | 28.0  | 28.0  | 28.0  |

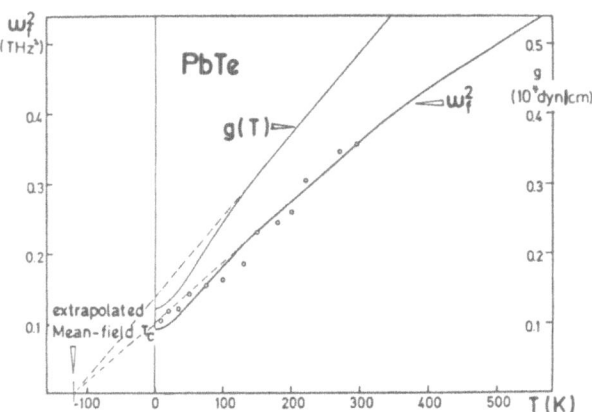

Fig. 3.5 Temperature dependence of the squared soft mode $\omega_f^2$ and the coreshell coupling constant g(T) of PbTe

It has to be pointed out, that in SnTe the experimentally observed dip in the TA mode at $qa = \frac{\pi}{3}$ cannot be fitted by the model, yet it is possible to explain and furthermore to describe this anomaly in terms of the coupling of the SPA-phonons to exact non-linear but periodic solutions of the model (Bilz et al., 1982; Bussmann-Holder et al., 1981a).

The temperature dependence of $\omega_f^2$ has been calculated for PbTe and SnTe. For the latter compound, the ferroelectric modes (ordinary and extraordinary) could also be fitted by means of the SPA.

Fig. 3.5 shows the temperature dependence of $\omega_f^2$ for PbTe. The dashed line indicates the meanfield regime, while the full line represents the results of the SPA-calculation. Obviously deviations from the $\gamma = 1$ behaviour occur for high as well as for small temperatures, the origin of which has been explained in the preceding section. It is interesting to note that, within the SPA, PbTe can never undergo a phase transition, as its harmonic electron-phonon-coupling term is repulsive, $g_2 > 0$, which indicates the stability of the system.

For SnTe we have calculated $\omega_f^2(T)$ in both regimes, the paraelectric and the ferroelectric one, and compared to experimental data. Deviations from mean-field behaviour show up in SnTe close to the actual phase transition point. While the extrapolation of the mean-field regime leads to a $T_c$ of 79 K, the calculated $T_c$ is 87 K which is in good agreement with the experiment (Fig. 3.6).

Because of the cubic structure in the paraelectric phase of SnTe the two transverse soft modes are degenerate. Below $T_c$ the crystals undergo a tetrahedral distortion. Consequently two soft modes are observed (Murase et al., 1979a). Within the model the splitting of the ferroelectric mode has been taken into account by a weighting factor of 3 for the extraordinary mode, that means $g_2^{(1)}$ (extra-ordinary) $= 3g_2^{(1)}$ (ordinary) and $g_4$ (extra-ordinary) $= 3g_4$ (ordinary), respectively. Fig.3.7 shows the comparison of experimental and theoretical data for the ferroelectric phase.

It is well known that defects in ferroelectric compounds lead to remarkable shifts of the soft mode frequency. For $SrTiO_3$ $O^{2-}$-defects induce a shift of $\omega_f$

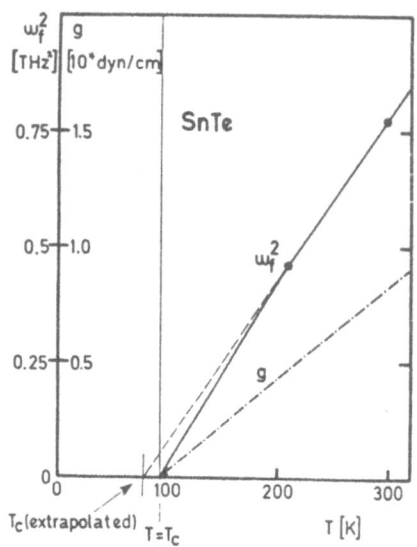

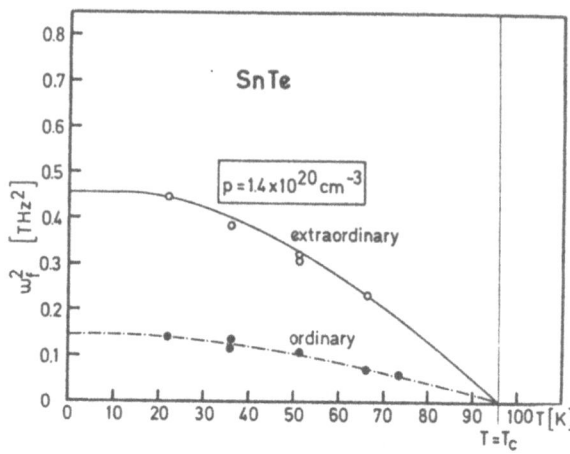

**Fig. 3.6** Temperature dependence of the squared soft mode $\omega_f^2$ and the coreshell coupling constant g(T) of SnTe in the paraelectric regime

**Fig. 3.7** Temperature dependence of the ordinary and the extraordinary soft modes of SnTe in the ferroelectric regime

of $\Delta\omega_f \cong 75,82$ cm$^{-1}$/atom % oxygen vacancies (Wagner et al., 1980; Bäuerle et al., 1980). Within the polarizability model the defects enter via changes in $g_2$ which means the Coulomb potential, while $g_4$ is not or only slightly affected by defects (Bussmann-Holder et al., 1981c). The harmonic part of the double well potential becomes more and more repulsive with increasing defect concentration and is finally positive, which is schematically visualized in Fig.3.8. For the IV-VI compounds it was assumed that precisely the same mechanism applies, $g_4$ remains constant, while $g_2$ is the only quantity which is affected. By means of this assumption the shift in $\omega_f^2$ could be calculated for varying doping concentrations and furthermore $T_c$ could be evaluated out of these data. Figures 3.9 and 3.10 show the results. The polarizability model predicts a linear dependence of $T_c$ on the doping rate which reproduces closely the experimental data. Yet the result of this calculation is contradictory to that by Kawamura (1977) who evaluates large deviations from linearity.

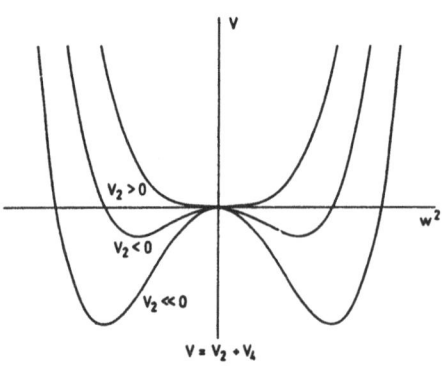

**Fig. 3.8** The potential energy V as a function of the squared relative core-shell displacement $w^2$ for three different values of its harmonic part $V_2$

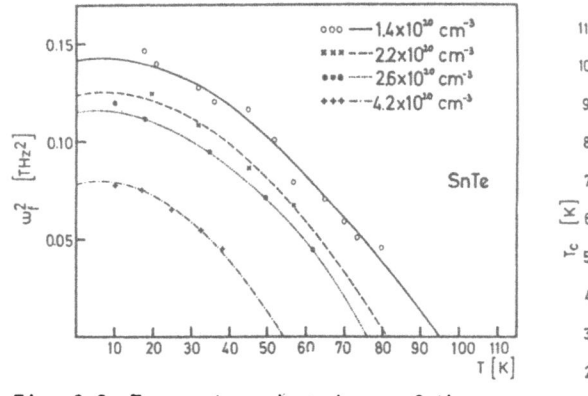

**Fig. 3.9** Temperature dependence of the squared soft mode $\omega_f^2$ of SnTe in the ferroelectric regime for various doping concentrations

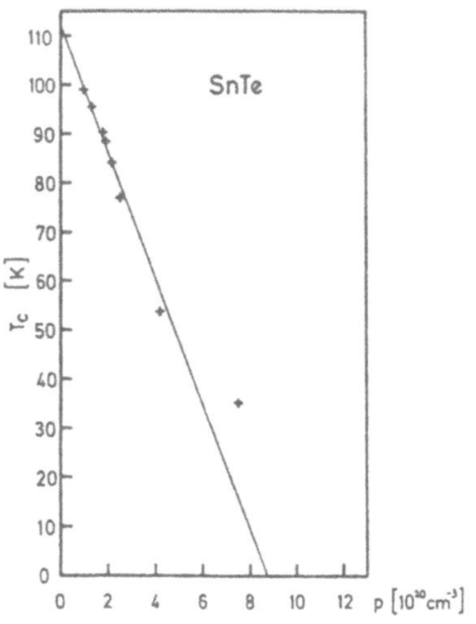

**Fig. 3.10** Phase transition temperature $T_C$ as a function of doping concentration

The change in $g_2$ is interpreted in terms of a decrease in the attractive Coulomb forces within the polarizable unit. The model does not admit to predict any influence of free carriers on the soft mode. But tight binding calculations by Vogl and Verges (1982) show that the free carriers leave the ferroelectric mode unchanged so that the observed shifts are nearly completely due to defects.

### 3.5.2 Three-Dimensional Models

The above outlined linear chain model gives a very transparent physical description of the ferroelectric soft mode behaviour in the (111)-direction. For other directions it has to be generalized and looses thereby most of its appealing simplicity.

For the description of the phonon dispersion curves in a general direction of the Brillouin zone a full shell model including long range Coulomb effects has to be used (Cochran et al., 1966). In the following it will be shown that in the (111)-direction this model is, to a good approximation, equivalent to the linear chain representation. The well known shell-model equations of motion read:

$$M\omega^2 u = [R + ZCZ - (T + ZCY)(S + K + YCY)^{-1}(T^+ + YCZ]u \quad . \quad (3.102)$$

R,T,S are the short range coupling matrices specifying the short range ion-ion, ion-shell and shell-shell interactions, respectively. The corresponding electrostatic interactions are given by ZCZ, ZCY and YCY. M and K are diagonal matrices characterizing the masses and the respective core-shell interactions of the ions. Since the (111) direction represents an alternating sequence of sheets carrying positive and negative ions, the Coulomb interaction ZCZ is basically reduced to

an effective nearest neighbour force constant between adjacent sheets and leads
therefore to a renormalization of the nearest neighbour force constants contained
in the coupling matrix R. The force constant for the longitudinal vibrations is ef-
fectively increased whereas the force constant for the transverse branches is ef-
fectively decreased. A similar renormalization occurs for the T and S contributions.
In the shell-shell coupling part we have to introduce further corrections which
arise from the fact that the different electronic polarizabilities of anions and
cations lead to different shell charges $Y_1$ and $Y_2$ and hence different prefactors
$Y_k Y_{k'}$, show up for the different $S_{kk'}$, components. In a good approximation we can
treat this effect by renormalizing the shell-shell coupling constants $g_2^{(1)}$ and $g_2^{(2)}$.
The force constants derived in this way from the best shell model fit of Cochran
et al. (1966) to the measured dispersion curves of PbTe (model VII) are compared to
the parameters of the linear chain model in Table 3.2. The first column gives the
force constants of Cochran for the transverse coupling in our notation. The second
column shows the renormalized force constants while in the third column the parameters obtained by fitting the linear model to the dispersion curves of the transverse branches in the (111)-direction are presented (Bussmann-Holder et al., 1980).
In general the agreement is quite good. The small differences may partly be due to
the different fitting procedures and partly to the approximations made in renormalizing the force constants of Cochran et al. (1966). In particular the results
prove that no additional changes have to be made in the force constants which
enter the T-matrix.

Table 3.2 The force constants of PbTe. Units are $10^4$ dyn/cm.
               (Colums 1) Force constants of Cochran for the transverse coupling;
               (Colums 2) renormalized force constants;
               (Colums 3) force constants of linear chain-model.

| | | | |
|---|---|---|---|
| f | 24.56 | 11.34 | 13.0 |
| f' | -0.13 | -0.137 | 0.75 |
| f" | -1.17 | -1.17 | -1.8 |
| h | 31.58 | 34.53 | 28.0 |
| g(T) | 6.85 | 0.30 | 0.45 |
| $y_1$ | -0.799 | - | - |
| $y_2$ | -3.703 | - | - |
| z | 2.15 | | |

In conclusion, it has to be mentioned that the polarizability model gives a uni-
fied picture of ferroelectricity in structurally completely different compounds.
It shows that ferroelectric soft modes can be described in terms of a non-linear
quartic polarizability of the chalcogenide ion. Furthermore the exact solutions
of this model (Bilz et al., 1982) which will be discussed later, provide a simple

picture to describe phonon anomalies, electrooptic phenomena and domain walls in terms of non-linear periodic excitations, "periodons" (Büttner and Bilz, 1981).

### 3.6 Comparison of Models

In the last section on the soft-mode behaviour of the IV-VI-compounds the three lattice dynamical models described above shall be compared to each other. A prerequisite for a phase-transition to occur is that long-range and short-range forces are of opposite sign and cancel each other at a certain temperature, the phase transition temperature $T_c$ (Cochran, 1960; Anderson, 1960). The stability of such a system in its high temperature phase is achieved via anharmonic interactions. This means for the required potential, that it has to be of double-well nature, with the harmonic part $V^{(2)}$ being negative and the anharmonic part $V^{(4)}$ being positive. This feature is common to all three models discussed before. A diagrammatic survey is given in Fig.3.11.

$$\omega_f^2(T) \quad = \quad \omega_{0f}^2 \quad + \quad 2\omega_{0f}\,\Delta_f(T)$$

| | SR | LR | $\|V_{st}^{(3)}\|^2/\Delta E$ | $\Sigma = \phi^{(2)}$ | $(\phi^{(3)})^2/\Delta E$ | $\phi^{(4)}$ | $V^{(4)}$ |
|---|---|---|---|---|---|---|---|
| 1. Anharmonic interaction model (Cowley, Bruce et al.) | >0 | <0 | - | <0 | + | + | - |
| 2. Vibronic model (Kristoffel, Konsin, Kawamura et al.) | >0 | <0 | <0 | <0 | - | + | - |
| 3. Polarizability model | >0 | <0 | - | <0 | - | - | + |

$\phi^{(4)} \hat{=}$ anharmonic ⟨lattice⟩ potential

$V^{(4)} \hat{=}$ ⟨local on-site⟩ potential (electron-two phonon coupling !!)

Fig. 3.11
Comparison of the different models

Quite generally the temperature dependence of the soft ferroelectric mode is given by:

$$\omega_f^2(T) = \omega_{0f}^2 + 2\omega_{0f}\Delta_f(T) \tag{3.103}$$

In the anharmonic lattice model $\omega_{0f}^2$ consists of a repulsive short-range term and an attractive long-range Coulomb term by means of which the harmonic frequency squared $\omega_{0f}^2$ becomes negative. The stabilization of the paraelectric phase is, in this model, achieved via a cubic term divided by the energy gap, $(\phi^{(3)})^2/\Delta E$ plus a fourth order lattice anharmonicity $\phi^{(4)}$, which introduces the temperature dependence of the soft mode.

In the vibronic model $\omega_{0f}^2$ consists again of the two parts cited above, the short-range repulsive term and the long-range attractive term. Yet in this model both terms together do not lead to a negative harmonic frequency $\omega_{of}^2$, which causes the lattice instability. In this model the harmonic frequency $\omega_{of}$ becomes imaginary because of the interband electron-phonon coupling term $(V_{st}^{(3)})^2/\Delta E$ which at the same time is temperature dependent. It has been shown that the T-dependence, which enters via the vibronic interaction term, is not sufficient to describe the observed temperature dependence of the soft mode. To overcome this difficulty a phonon-phonon anharmonicity has to be introduced which exhibits the same features as in the anharmonic lattice model.

In the polarizability model the harmonic core-shell force constant $g_2^{(1)}$ contains again a repulsive short-range force and an attractive Coulomb force by means of which $\omega_{of}^2$ becomes negative so that in this model there is no need to introduce an interband electron-phonon-coupling term. The harmonic instability of the polarizability model is, in the paraelectric phase, compensated by a fourth-order electron-ion coupling term $g_4$.

Neither a cubic term in the potential nor a phonon-phonon fourth-order term is needed to explain the observed experimental data and to get excellent agreement with experiment not only for the IV-VI-compounds but also for several other ferroelectric crystals. Thus the most important difference between the anharmonic-lattice, the vibronic and the polarizability model does not concern the harmonic frequency but the anharmonicity. While for both the anharmonic and the vibronic model an anharmonic intersite lattice potential $V^{(4)}$ is used to describe the temperature dependence of the soft mode, the polarizability model takes advantage of the local instability of the chalcogenide ions $X^{2-}$ ($X = 0$, S, Se,Te) and explains the soft-mode behaviour by means of a local on-site electron-two-phonon coupling term.

## 3.7 Nonlinear Excitations in IV-VI Semiconductors

In the preceding chapter the simple diatomic shell-model has been introduced to explain the observed soft-mode behaviour in the IV-VI compounds and related lattice dynamical properties. In this section the soft-mode-concept is extended to the non-linear case, where exact periodic solutions of the model are described. These exact solutions of a non-linear three-dimensional lattice are compared to the self-consistent phonon approximation and especially the coupling between these non-linear modes and the self-consistent phonons is discussed (Bilz et al., 1982). It is shown that this mode coupling leads to either ferroelectric-type phases or to commensurate phases and represents the origin of the experimentally observed phonon anomalies in crystals with phase transitions.

The harmonic lattice potential of ionic solids, the dynamics of which are described in terms of a dipolar shell model, is given by

$$\phi^{(2)} = \phi_{ii} + \phi_{ei} + \phi_{ee} , \qquad (3.104)$$

where the ion-ion interaction is given by

$$\phi_{ii} = \frac{1}{2} \sum_{LL'} u(L)\phi_i(L,L')u(L') \quad , \tag{3.105}$$

the electron-electron-interaction by

$$\phi_{ee} = \frac{1}{2} \sum_{LL'} v(L)\phi_e(L,L')v(L') \quad , \tag{3.106}$$

while the electron-ion interaction is:

$$\phi_{ei} = \frac{1}{2} \sum_{LL} [v(L)\phi_{ei}(L,L')u(L') + c.c.] \quad . \tag{3.107}$$

$u(L)$ and $v(L)$ are the displacements of the ions and electronic shells at lattice site $L = (\ell,\kappa)$, respectively.

An extension of the harmonic model is obtained by including a local fourth-order non-linear potential in the electron-ion interaction:

$$\phi_{ei}^{(4)} = \frac{1}{4} \sum_L w^2(L)\phi_{ei}^{(4)}(L)w^2(L) = \frac{1}{4} \sum_{L,\alpha} g_{4,\alpha}w^4(L) \quad , \tag{3.108}$$

with the difference coordinate $w(L) = v(L) - u(L)$ and $\alpha = x,y,z$. This anharmonic potential causes a non-linear polarizability which has been shown to be responsible for structural phase transitions in many ferroelectric systems. The equations of motion are given by:

$$M(\kappa) \ddot{u}(L) = \partial\phi/\partial u(L) \quad , \tag{3.109}$$

$$m_\ell(\kappa) \ddot{v}(L) = \partial\phi/\partial v(L) = 0 \quad , \tag{3.110}$$

with $M(\kappa)$ and $m(\kappa)$ being the ionic and effective electron-shell masses at site $\kappa$. For the electron-shell equations the adiabatic condition is used. The self-consistent-phonon approximation, as has been discussed in Sect. 3.5, corresponds to the following substitution:

$$w_\alpha^3(L) \cong 3 w_\alpha(L) <w_\alpha^2(L)>_T \quad , \tag{3.111}$$

where $<w_\alpha^2(L)>_T$ represents the self-consistent thermodynamical average at temperature T. Exact non-linear periodic lattice solutions are obtained from the following ansatz for the displacements $X(X = u,v,w)$:

$$X(L) = X_1 \exp\{i[\omega t - qR(L)]\} + X_3 \exp\{3i[\omega t - qR(L)]\} \quad . \tag{3.112}$$

$X_1$ and $X_3$ are determined by the non-linear and linear coupling constants, where

89

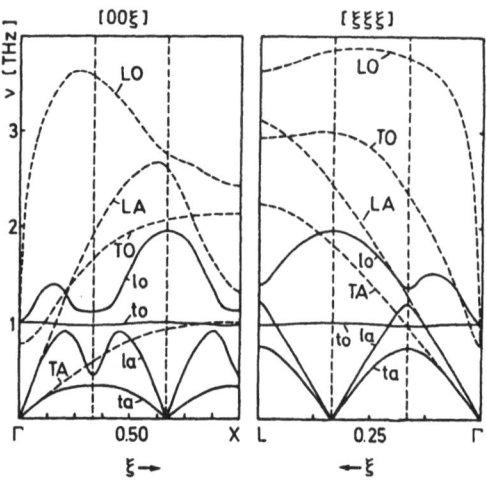

Fig. 3.12 Dispersion curves of phonons (dashed lines) and periodons (solid lines) in SeTe at 100K. Capital (small) letters denote polarization of phonons (periodons)

one has to take the real parts of Eq. (3.112). In particular, the amplitude $X_3$ vanishes for $w(L)$ if the non-linear coupling coefficient $\phi_{ei}^{(4)}(L) \neq 0$. The dispersion relations for the periodons $\omega_p(q)$ is given by

$$\omega_p(q) = \frac{1}{3}\,\omega(3q),\qquad\qquad\qquad\qquad (3.113)$$

where $\omega(q)$ is the dispersion relation in the SPA in the limit $\phi_{ei}(L,L)\to\infty$ for those lattice points $L$ with $\phi_{ei}^{(4)}(L) \neq 0$. It has to be pointed out that for optical branches the periodon amplitudes for certain wave vectors may become imaginary and consequently real solutions (besides the trivial ones with $X_1=0$) do not exist. In Fig. 3.12 phonons and periodons are shown for SnTe. Figure 3.13 contains the dispersion of the periodon amplitude. Of particular interest is the interaction between periodons and phonons which leads to the anomalous dispersion of the lowest acoustic branch in the (111)-direction in SnTe as shown in Fig. 3.14. To describe this

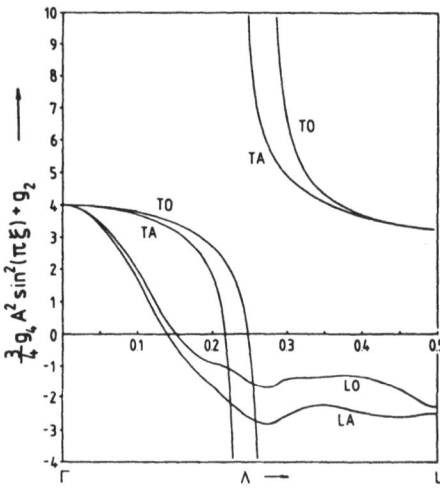

Fig. 3.13 Dispersion of the periodon amplitudes in the $\Lambda$-direction

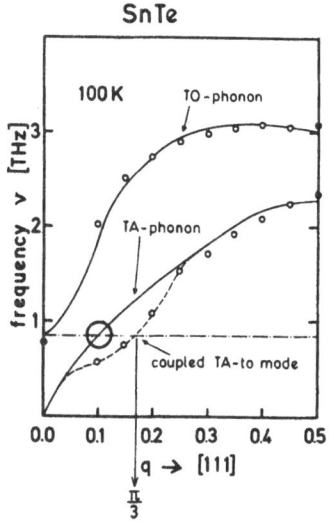

SnTe

100 K    TO-phonon

TA-phonon

coupled TA-to mode

frequency ν [THz]

0.0   0.1   0.2   0.3   0.4   0.5

q → [111]

$\frac{\pi}{3}$

Fig. 3.14  Dispersion of coupled phonon-periodon mode in the paraelectric regime of SnTe

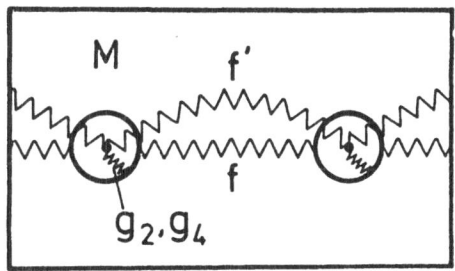

M        f'

f

$g_2, g_4$

Fig. 3.15  Monatomic linear chain model

interaction between phonons and periodons a simplified isotropic model is used consisting of one polarizable ion in the unit cell and nearest-neighbour inter-action only (Fig. 3.15), where only transverse solutions are considered. The re-sulting equations of motion are:

$$M \ddot{u}_n = g_2 w_n + g_4 w_n^3 + f'(u_{n+1} + u_{n-1} - 2u_n) \quad , \tag{3.114}$$

$$0 = -g_2 w_n - g_4 w_n^3 + f(w_{n+1} + w_{n-1} - 2w_n) + f(u_{n+1} + u_{n-1} - 2u_n). \tag{3.115}$$

The exact periodon solutions are:

$$w_n = A \sin (\omega t - nqa), \tag{3.116}$$

$$u_n = B \sin (\omega t - nqa) + C \sin 3(\omega t - nqa), \tag{3.117}$$

with the dispersion relation

$$M \omega_p^2 = \frac{4}{3} (f + f') \sin^2 \left(\frac{3}{2} qa\right) \quad , \tag{3.118}$$

and the amplitude

$$A^2 = -\frac{4g_2}{3g_4} \left[ 1 + \frac{4f}{g_2} \frac{M \omega_p^2 - 4f' \sin^2 \frac{1}{2} qa}{M \omega_p^2 - 4(f+f') \sin^2 \frac{1}{2} qa} \sin^2 \frac{1}{2} qa \right] \quad . \tag{3.119}$$

The amplitudes B and C have a similar structure and can be expressed as functions of A. The dispersion relation which corresponds to the SPA is

91

$$M \, \omega_s^2 = 4 \sin^2 \tfrac{1}{2} \, qa \left[ \frac{f}{1 + \frac{4f}{g(T)} \sin^2 \tfrac{1}{2} \, qa} + f' \right] \qquad (3.120)$$

with $g(T) = g_2 + 3g_4 <w_n^2>_T$.

The coupling between periodons and phonons is described by the ansatz:

$$w_n = w_{np} + w_{ns} \quad , \qquad (3.121)$$

$$u_n = u_{np} + u_{ns} \quad , \qquad (3.122)$$

which splits the equation of motion into two sets, one in $u_{np}$, $w_{np}$ and the other in $u_{ns}$, $w_{ns}$. The index (np) denotes the periodon-part and the index (ns) the phonon part of the displacements, respectively.

$$M \, \ddot{u}_{ns} = (f + f') \, \{u_{ns+1} + u_{ns-1} - 2u_{ns}\} + f \, \{w_{ns+1} + w_{ns-1} - 2 \, w_{ns}\} \quad , \qquad (3.123)$$

$$0 = (g_2 + 3g_4 \, w_{np}^2) \, w_{ns} + g_4 \, w_{ns}^3 - f(w_{ns+1} + w_{ns-1} - 2w_{ns})$$

$$- f(u_{ns+1} + u_{ns-1} - 2u_{ns}) \quad , \qquad (3.124)$$

$$M \, \ddot{u}_{np} = (f + f') \, \{u_{np+1} + u_{np-1} - 2u_{np}\} + f(w_{np+1} + w_{np-1} - 2w_{np}) \quad . \qquad (3.125)$$

$$0 = (g_2 + 3g_4 \, w_{ns}^2) \, w_{np} + g_4 \, w_{np}^3 - f \, (w_{np+1} + w_{np-1} - 2w_{np}) . \qquad (3.126)$$

This leads to a renormalized harmonic core-shell force constant $g_2$, which for the self-consistent solutions now becomes

$$g_2 \rightarrow g_2 + 3 \, g_4 \, w_{np}^2 = \tilde{g}_2 \quad , \qquad (3.127)$$

and for the periodon solutions

$$g_2 \rightarrow g_2 + 3 \, g_4 w_{ns}^2 = g(T). \qquad (3.128)$$

This means that via the harmonic electron-ion-coupling the coupling to the non-linear solutions enters the self-consistent solutions and that by means of the same coupling mechanism the periodon amplitudes become temperature dependent. Two different temperature regimes have to be distinguished now: The high-temperature paraelectric regime with self-consistent phonons in a fluctuating periodon field, and the low-temperature regime which is governed by static periodons, which leads to a site-dependent electron-ion coupling

$$g(T) + 3g_4 \, w_{np}^2 \left( \frac{2\pi}{3} \right) = \begin{cases} g(T) & n = 0 \\ -2g(T) - \frac{9f'f}{f'+f} & n = 1,2 \end{cases} \qquad (3.129)$$

and causes a tripling of the lattice constant. For high temperatures the fluctuating periodons are approximated by a time-averaged solution which leads to a q-
dependent coupling:

$$g(T) + 3g_4 \, w_{np}^2(q) = g(T) + 3g_4 \, A_T^2(q) \, \langle \sin^2(\omega t - nqa) \rangle_{time} \qquad (3.130)$$

$$= g(T) + \frac{3}{2} \, g_4 \, A_T^2(q) \quad ,$$

with $A_T^2(q)$ being the periodon amplitude squared and $g(T)$ the temperature dependent
renormalized core-shell coupling constant. By means of this periodon-phonon coupling
it is possible to describe as well phase transitions of inhomogeneous nature as
has been observed in $K_2SeO_4$ (Fig.3.16) as the observed phonon anomalies and the
transition in IV-VI compounds (Fig. 3.14). Within the above described model $g(T)$
determines the temperature dependence of a phase transition to a commensurate phase.

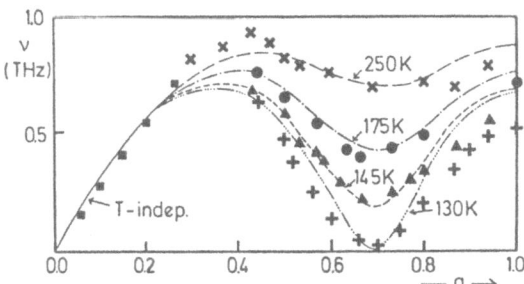

Fig. 3.16. Dispersion of coupled
phonon-periodon mode in the paraelectric regime ($T > T_i$) of $K_2SeO_4$

In a diatomic lattice, the coupling between phonons and periodons exhibits matrix
character, i.e. acoustic-acoustic, acoustic-optic, optic-acoustic and optic-optic
couplings occur. For SnTe, quantitative calculations (Bussmann-Holder, unpublished)
reproduce the phonon anomaly near $qa=\pi/3$ (see Fig. 3.14) in terms of an optic peri-
odon-acoustic phonon coupling. This anomaly indicates an incipient commensurate structural instability of the SnTe lattice in addition of the ferroelectric phase transition.

To describe transitions to incommensurate structures it is necessary to include
the temperature dependence of $f'$, the nearest neighbour core-core coupling, which
shifts the minimum of $\omega(q)$ away from a commensurate value.

In conclusion,it has been shown that systems with structural phase transitions
may be treated in terms of coupled modes, one of which is a self-consistent phonon
while the other is a periodon, i.e. a non-linear periodic lattice wave.


## 4. Summary and Conclusion

The group-IV chalcogenides exhibit a number of unusual and interesting properties.
They have small energy gaps, are electronically highly polarizable and show—as
one of their most outstanding properties—a soft-mode behaviour and dielectric

anomalies which lead to the conclusion that they are either real ferroelectrics or incipient ones. Since the discovery of ferroelectricity several theories tried to explain this phenomenon. Besides the phenomenological concept of Landau concerning structural phase transitions (which has first been applied to ferroelectrics by Ginzburg (1945,1949) and Devonshire (1949,1951,1954)) several microscopic and semimicroscopic theories have been developed. Cochran et al. (1966) was the first who suggested that the IV-VI semiconductors belong to the class of ferroelectric systems. This theory is based on the assumption that the compensation of long-range and short-range forces which leads to a ferroelectric phase transition is caused by a phonon-phonon anharmonicity. Kristoffel et al. (1967,1968,1969,1971,1972) and Bersuker and Vekhter (1967,1969a,1969b) assume that the ferroelectric phase transition in IV-VI compounds originates in the interband electron-phonon interaction. In these models, the destabilization is caused by a third-order electron-phonon coupling. Similar to Cochran (1960), the fourth-order ion-ion interaction stabilizes the para- and ferroelectric phase. The polarizability model proposed by Migoni et al. (1976) is based on the local non-linear electron-ion interaction of the chalcogenide ions which provides the driving mechanism for ferroelectricity in this model. So far this theory is the only one which is able to reproduce quantitatively the complex temperature dependence of soft modes in many different ferroelectrics (Bilz et al., 1979; Bussmann-Holder et al., 1981) and the Raman spectrum of ferroelectrics ($KTaO_3$). The same model provides exact non-linear solutions which lead to anomalies in the dispersion curves of the IV-VI compounds when they are coupled to the acoustic mode. The model is consistent with the change of soft-mode frequencies under the influence of lattice defects.

The authors are indebted to W. Kress for a critical reading of the manuscript.

## References

Abbati, I., Braikovich, L., Cuicci, G. (1977): Nuovo Cimento **39** B, 727
Albers, W., Haas, C., Ober, H., Schodder, G.R., Wasscher, J.D. (1962): J. Phys. Chem. Sol. **23**, 215
Anderson, P.W. (1960): *Fizika Dielectrikov*, ed. by G.J. Skanavi (Akad. Nauk. SSSR, Moskow)
Animalu, A.O.E., Heine, V. (1966): In *Pseudopotentials in the Theory of Metals*, ed. by W.A. Harrison (Benjamin, New York)
Bate, R.T., Carter, D.L., Wrobel, J.S. (1970): Proc. 10th Internat. Conf. Phys. Semiconductors, ed. by S.P. Keller, J.C. Hensel, F. Stern (Nat. Bureau of Standards, Springfield) p. 125
Bäuerle, D., Wagner, D., Wöhlecke, M., Dorner, B., Kraxenberger, H. (1980): Z. Phys. B **38**, 335
Balkanski, M., Teng, M.K., Massot, M., Bilz, H. (1980): Ferroelectrics **26**, 737
Bantle, W. (1942): Helv. Phys. Acta **15**, 373
Barrett, G.H. (1952): Phys. Rev. **86**, 118
Bernick, R.L., Kleinmann, L. (1970): Solid State Commun. **8**, 569
Bersuker, I.B., Vekhter, B.G. (1967): Fiz. Tverd. Tela **9**, 2652
Bersuker, I.B., Vekhter, B.G. (1969a): Fiz. Tverd. Tela **9**, 2652
Bersuker, I.B., Vekhter, B.G. (1969b): Izv. Akad. Nauk, SSSR Ser. Fiz. **33**, 199

Bilz, H., Bussmann, A., Benedek, G., Büttner, H., Strauch, D. (1980): Ferroelectrics **25**, 339

Bilz, H., Büttner, H., Bussmann-Holder, A., Kress, W., Schröder, U. (1982): Phys. Rev. Lett. **48**, 264

Blinc, R., Zeks, B. (1974): *Soft Modes in Ferroelectrics and Antiferroelectrics* (Elsevier, New York)

Born, M., Huang, K. (1954): *The Dynamical Theory of Crystal Lattices* (Oxford University Press, Oxford)

Boyer, L.L. (1981): Ferroelectrics **35**, 83

Büttner, H., Bilz, H. (1981): In *Recent Developments in Condensed Matter Physics*, Vol.1, ed. by J.T. Devreese (Plenum, New York) p. 49

Burke, J.R., Houston, B., Savage, H.T. (1970): Phys. Rev. B **2**, 1977

Burstein, E., Pinczuk, A., Wallis, R.F. (1971): In *The Physics of Semimetals and Narrow-Gap-Semiconductors*, ed. by D.L. Carter, R.T. Bate (Pergamon, Oxford) p. 251

Busch, G. (1938): Helv. Phys. Acta **11**, 269

Busch, G., Scherrer, P. (1935): Naturwissenschaft **23**, 737

Bussmann-Holder, A., Bilz, H., Roenspiess, R., Schwarz, K. (1980a): Ferroelectrics **25**, 343

Bussmann-Holder, A., Bilz, H., Kress, W. (1980b): J. Phys. Soc. Jpn. **49** A, 737

Bussmann-Holder, A., Bilz, H., Kress, W., Schröder, U., (1981a): In *Physics of Narrow Gap Semiconductors*, ed. by E. Gornik, H. Heinrich, L. Palmetshofer (Springer, Berlin, Heidelberg, New York) p. 257

Bussmann-Holder, A., Benedek, G., Bilz, H., Mokross, B. (1981b): J. Phys. C**6**, C6-409

Bussmann-Holder, A., Bilz, H., Bäuerle, D., Wagner, D. (1981c): Z. Phys. B**41**, 353

Car, R., Ciucci, G., Quartapelle, L. (1978): Phys. Stat. Sol. B**86**, 471

Chandrasekhar, H.R., Zwick, U. (1976): Solid State Commun. **18**, 1509

Chandrasekhar, H.R., Humphreys, R.G., Zwick, U., Cardona, M. (1977): Phys. Rev. B**15**, 2177

Chattopadhyay, T., Werner, A., v. Schnering, H.G. (1983): AIRAPT Conference, Albany, N.Y. (to be published)

Cochran, W. (1960): Adv. Phys. **9**, 387

Cochran, W., Cowley, R.A., Dolling, G., Elcombe, M.M. (1966): Proc. R. Soc. A**932**, 433

Cohen, M.L., Heine, V. (1970): Solid State Phys. **24**, 37

Cowley, R.A. (1965): Philos. Mag. **11**, 673

Cuff, K.F., Ellett, M.R., Kuglin, C.D., Williams, L.R. (1964): Proc. 7th Int. Conf. Physics Semiconductors, ed. by M. Hulin (Dunod, Paris) p. 677

Cuff, K.F., Ellett, M.R., Kuglin, C.D. (1961): J. Appl. Phys. Suppl. **32**, 2169

Cuff, K.F., Ellett, M.R., Kuglin, C.D. (1962): Proc. 6th Int. Conf. Physics Semiconductors (Institute of Physics and the Physical Society, London)

Dalven, R. (1973): Solid State Phys. **28**, 179

Devonshire, A.F. (1949): Philos. Mag. **40**, 1040

Devonshire, A.F. (1951): Philos. Mag. **42**, 1065

Devonshire, A.F. (1954): Adv. Phys. **3**, 85

Duggin, M.J. (1972): J. Phys. Chem. Solids **33**, 1029

Eisenriegler, E. (1974): Phys. Rev. B **9**, 1029

Fischer, C.F. (1972): Atomic Data **4**, 301

Fröhlich, H. (1949): *Theory of Dielectrics* (Clarendon, Oxford)

Goldak, J., Barrett, C.S., Innes, D., Youdelis, W. (1966): J. Chem. Phys. **44**, 3323

Grandke, T., Ley, L. (1977): Phys. Rev. B**16**, 832

Grandke, T., Ley, L., Cardona, M. (1978): Phys. Rev. B**18**, 3847

Gillis, N.S., Koehler, T.R. (1972): Phys. Rev. B**5**, 1924

Ginzburg, V.L. (1945): Zh. Eksp. teor. Fiz. **15**, 739

Ginzburg, V.L. (1949): Zh. Eksp. teor. Fiz. **19**, 36

Harman, T.C. (1971): In *Physics of IV-VI Compounds and Alloys*, ed. by S. Rabii (Gordon and Breach, London) p. 141

Harrison, W.A. (1980): *Electronic Structure and the Properties of Solids* (Freeman, San Francisco)

Heidrich, K., Staude, W., Treusch, J., Overhof, H. (1974): Phys. Rev. Lett. **33**, 1220

Heidrich, K., Staude, W., Treusch, J., Overhof, H. (1975): Solid State Commun. **16**, 1043

Herman, F., Kortum, R.L., Ortenberg, I.B., van Dyke, J.P. (1968): J. Phys. (Paris) **29**, C4

Hohnke, D.K., Holloway, H., Kaiser, S. (1972): J. Phys. Chem. Solids **33**, 2053

Iizumi, M., Hamaguchi, Y., Komatsubara, K.F., Kato, Y. (1975): J. Phys. Soc. Jpn. **38**, 443

Katayama, S., Mills, D.L. (1980): Phys. Rev. B**22**, 336

Kawamura, H., Murase, K., Sugai, S., Takaoka, S., Nishikawa, S., Nishi, S., Katayama, S. (1977): Proc. Int. Conf. Lattice Dynamics, ed. by M. Balkanski (Flammarion Sciences, Paris) p. 658

Kawamura, H. (1977): In *Physics of Narrow Gap Semiconductors*, ed. by J. Rautuszkiewicz, M. Gorska, E. Kacmarek (Elsevier, New York) p. 7

Kemeny, P.C., Cardona, M. (1976): J. Phys. C**9**, 1361

Kemeny, P.C., Azoulay, J., Cardona, M., Ley, L. (1977): Nuovo Cimento, **39**B, 709

Kirsch, R., Gérard, A., Wautelet, M. (1974): J. Phys. C**7**, 3633

Kohn, S.E., Yu, P.Y., Petroff, Y., Shen, Y.R., Tsang, Y., Cohen, M.L. (1973): Phys. Rev. B**8**, 1477

Kobayashi, K.L., Kato, Y., Katayama, Y., Komatsubara, K.F. (1966): Phys. Rev. Lett **37**, 160

Kobayashi, K.L., Katayama, Y., Narita, K., Komatsubara, K.F. (1979): Proc. 14th Int. Conf. Phys. Semiconductors, Inst. Phys. Conf. Ser. **43**, 437

Kowalczyk, S.P., Ley, L., McFeely, F.R., Shirley, D.A. (1974): J. Chem. Phys. **61**, 2850

Kristoffel, N.N., Konsin, P.I. (1967): Izv. Akad. Nauk. Est. SSR. Ser. fiz. math. **16**, 429

Kristoffel, N.N., Konsin, P.I. (1968): Phys. Stat. Sol. **28**, 731

Kristoffel, N.N., Konsin, P.I. (1969): Izv. Akad. Nauk. SSR. Ser. fiz. math. **18**, 439

Kristoffel, N.N., Konsin, P.I. (1971): Fiz. Tverd. Tela **13**, 3513

Kristoffel, N.N., Konsin, P.I. (1972): Sov. Phys. Solid Stat. **13**, 2969

Kunc, K., Martin, R.M. (1982): Phys. Rev. Lett. **48**, 406

Kurchatov, I.V. (1933) Segnetoelektriki Moskow: C-M 6111

Kwok, P.C., Miller, P.B. (1966): Phys. Rev. **151**, 387

Lambros, A.P., Geraleas, D., Economou, N.A. (1974): J. Phys. Chem. Solids **35**, 537

Levine, B.F. (1973): J. Chem. Phys. **59**, 1463

Ley, L., Cardona, M., Pollak, R.A. (1979): In *Photoemission in Solids*, Topics in Applied Physics, Vol.27, ed. by L. Ley, M. Cardona (Springer, Berlin, Heidelberg, New York) p. 11

Lin, P.J., Kleinman, L. (1966): Phys. Rev. **142**, 478

Littlewood, P.B. (1980): J. Phys. C**13**, 4855, 4875

Littlewood, P.B. (1982): In *Physics of Narrow Gap Semiconductors*, ed. by E. Gornik, H. Heinrich, L. Palmetshofer (Springer, Berlin, Heidelberg, New York) p. 238

Lovett, D.R. (1977): *Semimetals and Narrow-Bandgap Semiconductors* (Pion Limited, London)

Lowndes, R.P., Martin, D.R. (1969): Proc. R. Soc. A**308**, 473

Lowndes, R.P. (1972): Phys. Rev. B**6**, 1490

Mariano, A.N., Chopra, K.L. (1967): Appl. Phys. Lett. **10**, 282

Martinez, G., Schlüter, M., Cohen, M.L. (1975): Phys. Rev. B**11**, 651

Matthias, B.T. (1949): Phys. Rev. **75**, 1771

Matthias, B.T., Remeika, J.P. (1949): Phys. Rev. **76**, 1886

Melvin, J.S., Hendry, D.C. (1979): J. Phys. C**12**, 3003

Migoni, R., Bilz, H., Bäuerle, D. (1976): Phys. Rev. Lett. **37**, 1155

Mitchell, D.L., Palik, E.D., Zemel, J.N. (1964): Proc. 7th Int. Conf. Physics Semiconductors, ed. by M. Hulin (Dunod, Paris) p. 325

Moldovonova, M., Sumitrova, St., Decheva, St. (1964): Fiz. Tverd. Tela **6**, 3717 [Soviet Phsics Solid State **7**, 2032 (1966)]

Müller, K.A., Burkhard, H. (1979): Phys. Rev. B**19**, 3593

Mula, G. (1982): In *Physics of Narrow Gap Semiconductors*, ed. by E. Gornik, H. Heinrich, L. Palmetshofer (Springer, Berlin, Heidelberg, New York) p. 252

Muldawer, L.J. (1972): J. Nonmetals **1**, 193

Murase, K., Sugai, S., Takaoka, S., Katayama, S. (1976): Proc. 13th Int. Conf. Physics Semiconductors, ed. by F.G. Fumi (North Holland, Amsterdam) p. 305

Murase, K., Sugai, S. (1979): Solid State Commun. **32**, 89
Murase, K., Sugai, S., Higuchi, T., Takaoka, S., Fukunaga, T., Kawamura, H.
 (1978): Proc. 14th Int. Conf. Physics, Semiconductors, Inst. Phys. Conf. Ser.
 **43**, 437
Murase, K. (1980): J. Phys. Soc. Jpn. **49**, Suppl. 725
Nakanishi, A., Matsubara, T. (1980): Progr. Theor. Phys. **63**, 1
Nii, R. (1963): J. Phys. Soc. Jpn. **18**, 456
Nii, R. (1964): J. Phys. Soc. Jpn. **19**, 58
Nishi, S., Kawamura, H., Murase, K. (1980): Phys. Status Solidi (b) **97**, 581
Parke, A.W., Srivastava, G.P. (1980): Phys. Status Solidi (b) **101**, K31
Patel, C.K.N., Slusher, R.E. (1969): Phys. Rev. **117**, 1200
Pauling, L. (1960): *The Nature of the Chemical Bond* (Cornell Univ. Press,
 Ithaca, New York)
Pawley, G.S., Cochran, W., Cowley, R.A., Dolling, G. (1966): Phys. Rev. Lett.
 **17**, 753
Pawley, G.S. (1968): J. Physique Suppl. **29**, C4-145
Penn, D.R. (1961): Phys. Rev. **128**, 2093
Polatoglou, H.M., Theodorou, G., Economou, N.A. (1982): In *Physics of Narrow
 Gap Semiconductors*, ed. by E. Gornik, H. Heinrich, L. Palmetshofer (Springer,
 Berlin, Heidelberg, New York) p. 221
Porod, W., Vogl, P. Bauer, G. (1980): J. Phys. Soc. Jpn. **49**, Suppl. A, 649
Porod, W., Vogl, P. (1982): In *Physics of Narrow Gap Semiconductors*, ed. by
 E. Gornik, H. Heinrich, L. Palmetshofer (Springer, Berlin, Heidelberg, New
 York) p. 247
Preier, H. (1979): Appl. Phys. **20**, 189
Pytte, E. (1972): Phys. Rev. B**5**, 3758
Rabii, S., Lasseter, R.H. (1974): Phys. Rev. Lett. **33**, 703
Ravich, Yu.I., Efimova, B.A., Smirnov, I.A. (1970): *Semiconducting Lead Chal-
 cagerides*, ed. by L.S. Stil'bans (Plenum, New York)
Rytz, D., Höchli, U.T., Bilz, H. (1980): Phys. Rev. B**22**, 359
Shalvoy, R.B., Fisher, G.B., Stiles, P.J. (1977): Phys. Rev. B**15**, 1
Sham, L.J. (1974): In *Dynamical Properties of Solids*, Vol.I, ed. by G.K. Horton,
 A.A. Maradudin (North-Holland, Amsterdam) p. 301
Shevchik, N.J., Tejeda, J., Langer, D.W., Cardona, M. (1973): Phys. Status Solidi
 (b) **57**, 245
Shirane, G., Hoshino, S., Suzuki, K. (1950): Phys. Rev. **80**, 1105
Siapkas, D.I., Kyriakos, D.S., Economou, N.A. (1976): Solid State Commun. **19**, 765
Silverman, B.D. (1964): Phys. Rev. **131**, 2478
Silverman, B.D. (1964): Phys. Rev. **135**, A 1596
Silverman, B.D., Joseph, R.I. (1963): Phys. Rev. **129**, 2062
Silverman, B.D. (1964): Phys. Rev. **133**, A207
Slater, J.C. (1941): J. Chem. Phys. **9**, 16
Slater, J.C. (1950): Phys. Rev. **78**, 748
Sugai, S., Murase, K., Katayama, S., Takaoka, S., Nishi, S., Kawamura, H. (1977):
 Solid State Commun. **24**, 407
Suski, T., Karpinski, J., Kobayashi, K.L.I., Komatsubara, K.F. (1981): J. Phys.
 Chem. Solids **42**, 479
Schiferl, D. (1974): Phys. Rev. B**10**, 3316
Schmeltzer, D. (1983): Phys. Rev. Lett. (to be published)
Schneider, T., Beck, H., Stoll, E. (1976): Phys. Rev. B**13**, 1123
Schubert, K., Fricke, H. (1951): Z. Naturforschung **6a**, 781; Structure Repts. **15**,
 72
Stegmeier, E.F., Harbeke, G. (1970): Solid State Commun. **8**, 1275
Stiles, P.J., Brodsky, M.H. (1972): Solid State Commun. **11**, 1063
Stiles, P.J., Burstein, E., Langenberg, D.N. (1962): Proc. Int. Conf. Physics
 Semiconductors, Exeter
Takano, S., Hotta, S., Kawamura, H., Kato, Y., Kobayashi, K.L.I., Kamatsubara,
 K.F. (1974): J. Phys. Soc. Jpn. **37**, 1007
Tanaka, H., Morita, A. (1979): J. Phys. Soc. Jpn. **46**, 523
Tessmann, G.R., Kohn, A.H., Shockley, W. (1953): Phys. Rev. **92**, 890
Thomas, H. (1969): IEEE Trans. Magn. **5**, 874

Thomas, H. (1971): *Structural Phase Transition and Soft Modes*, ed. by Samuelsen, E.J. (Universitetsvorleiget, Oslo) p. 15

Thomas, H., Müller, K.A. (1972): Phys. Rev. Lett. **28**, 820

Thorhallson, G., Fisk, C., Fraga, S. (1968): Theoret. Chem. Acta **10**, 388

Tung, Y.W., Cohen, M.L. (1969): Phys. Rev. **180**, 823

Tung, Y.W. (1970): Phys. Rev. B**2**, 1216

Valasek, J. (1920): Phys. Rev. **15**, 537

Valasek, J. (1921): Phys. Rev. **17**, 475

Vlachos, S.V., Lambros, A.P., Economov, N.A. (1976): Solid State Commun. **19**, 759

Vogl, P., Verĝes, J.A.: To be published

Wagner, D., Bäuerle, D., Schwabl, F., Dorner, B., Kraxenberger, H. (1980): Z. Phys. B**37**, 317

Wagner, G.W., Thompson, A.G., Willardson, R.R. (1971): J. Appl. Phys. **42**, 2515

Warschauer, D. (1963): J. Appl. Phys. **34**, 1853

Watson, R.E. (1958): Phys. Rev. **111**, 1108

Wiley, J.D., Breitschwerdt, A., Schönherr, E. (1975): Solid State Commun. **17**, 355

Wiley, J.D., Buckel, W.J., Schmidt, R.L. (1976): Phys. Rev. B**13**, 2489

Wul, B., Goldman, I.M. (1945): C.R. Acad. Sci. URSS **46**, 139; **49**, 177

Wul, B., Goldman, I.M. (1946): C.R. Acad. Sci. URSS **51**, 21

Yin, M.T., Cohen, M.L. (1982a): Phys. Rev. B**26**, 3259

Yin, M.T., Cohen, M.L. (1982b): Phys. Rev. B**26**, 5668

# Combined Subject Index

$ABO_3$-compounds  52
Absorption coefficient  7,8
- edge  29
-, optical  57
Anharmonic effects  68,70,72,88,94
- interactions  52,65,87
- lattice model  53,71,87,88
Anisotropy  37,75
-, effective mass  12
Atomic electronegativity scale  56
Attenuation measurements  31

Band inversion  60
Barrett's formula  80
$BaTiO_3$  52
Birefrigence  29,30
Black phosphorus  56
Burn's model  43
Burstein-Moss shift  40

Carrier concentration  3,7,8,10,12,17,
  21,23,27,28,30,37,39,40
-, scattering of free  30
Central peak  43,44
Chemical trends  36,37,38
Coercive field  51
Coloumb forces  52,61,75,85,88
Commensurate phases  88,93
Covalency  38,56,61
Critical exponents  36,79,80
Critical temperature  2-4,8-10,17,18,
  20,24,29,30,32,34,36,39,44,51

Critical wave vector  67,68
Curie temperature  25,38,51,78
Curie-Weiss law  2,18,21,23,32,35,67
Cyclotron resonance  19,20,26,27,29

Defects  5,20,24,25,39,41,43,83,84
Deuteration  52
Dielectric anomaly  17,26,27,51,52,94
- constant  12,15,19-22,24,25,28,41-
  44,60,64,80
- function  11-14,18-20,25-27,38,42,
  62,63
- properties  4,24,25,28,29,39,51
Differential capacitance measurement
  20,21,23,24,36
Dipolar forces  37,51
Domain walls  8,19,29,30
Double well potential  52,84

Effective mass  12,16,26,27
Electrical resistivity  4,30,59
Electron-ion interaction  89,94
Electron-phonon interaction  7,28,35,
  37,38,40,44,71,73,88
Electronic energy bands  25,29,54,57
Electrooptic phenomena  87

Fabry-Perot interference  14,27
Faraday configuration  26,29
Far-infrared spectroscopy  5,8,11,17,
  18,24,26,29,36,41
Ferroelectricity  2,4,52-54,60,61,63,
  86,94

Ferroelectric domains  8,19,30
- materials  51,61,75,76,78,79,86,89
- mode  74,77-79,81,83,85-87
- phase transition  4,5,34,41,45,52,
    57,60,67,71,74,94
Ferromagnetic properties  51
Free carriers  2,5,8-10,12,15,17,23,
    25,26,35,39,42,43,73,85
Free energy  34,53,64,67,68,71

GaAs  62
GeTe  54
Group V elements  54,56

Hall effect  21,27,59
Hydrogen bonds  51,52
Hydrostatic pressure  30,40,56,59

Incommensurate phases  67,93
Inelastic tunneling  5,7
Ionicity  37,56,61

Jahn-Teller effect  38

KDP-type ferroelectrics  51
Kramers-Kronig relation  12
$K_2SeO_4$  74

Landau theory  31,34,35,52,53,77,80
Lattice defects  9,23,39,94
Linear-chain model  75,85
Lyddane-Sachs-Teller relation  2,12,
    24,28,30,36,41,60

Magnetic field  25-27,29,30,37,44,59
Magnetooptical experiments  1,16,27
Magnetoplasma effects  25,26,28,44
Mean-field approximation  35,36,60,70,
    79,83
Microscopic theories  60
Microwave techniques  25,28

Non-linear periodic, lattice waves
    76,83,87,89,93,94
Non-linear polarizability  74,86,89

Optical properties  20,25,26,29
Order parameter  5,53,68,80
Oscillator model  12
Oxygen  74

Paraelectric phase  52,54,67,69,73,75,
    77,81,87,88,92,94
Pauli-force model potential  38
Penn's model  37
Periodons  87,90-93
Perovskites  52,64,74
Phase transitions  3,5,15,18,29,31,36,
    60,61,65,68,71,73,75,78,87,88,93
Phase transition temperature  23,39,53,
    60,65,70,72,73,79,81,83,87
Phonon anomalies  87,88,93
Phonon-phonon anharmonicity  73,88,94
Photoemission measurements  57
Plasma edge  19,20,26
Plasma frequency  11-13,15,16
Plasmon-phonon mode  13,19
p-n junctions  2,20,21
Polarity  56,61
Polarizability  23,25,35,37,62,74,76,
    82,86
- model  53,74,79-81,84,86,88,94
Polarization  18,21,23,41,51,53,54,71
Pseudopotential  38,45,56,57,60,61,62,
    63

Quantum chemistry methods  56
- ferroelectrics  32,33,36,80
Quasi-harmonic approximation  52,65,
    67,68

Raman scattering  3,7,8,10,11,18,20,
    32,35,43,57,94
Reflectivity, infrared  11-14,16,17,
    19,27
Refractive index  26,27,29,30
Renormalization group theory  80
Resistance, anomaly  30
Rochelle salt  51

SbSI  74

Schottky barriers  2,20,21,24,36

Self-consistent phonons  35,36,70,76, 77,80,83,88,92,93

Shell model  35,74,85,86,88

Short-range potentials  37,52

Si  62

Soft modes  2-4,11,14,18,32,52,60,63, 64,87

Space charges  21,22,23

Spin-orbit interaction  57

Structural phase transition  36,37,39, 51-53,89,93,94

Superconductivity  64

Surface states  8

Szigeti model  37,41

Tight-binding model  56

Transmission  11,13,27,41

Ultrasonic wave velocities  31

Vacancies  24,40

Vibronic model  53,71,73,87,88

Watson sphere  74

X-ray analysis  3,10,23,31,57

Zero gap  1,28,59,60

# Springer Series in Solid-State Sciences

Editors: M. Cardona, P. Fulde, H.-J. Queisser

The series is devoted to single- and multi-author graduate-level monographs and textbooks in the areas of solid-state physics, solid-state chemistry, and solid-state technology. Proceedings of topical conferences which delineate the directions for significant future research are included, as are books on non-linear phenomena in solid-state physics.

Volume 40
K. Seeger
## Semiconductor Physics
An Introduction
2nd corrected and updated edition. 1982.
288 figures. XII, 462 pages
ISBN 3-540-11421-1

Volume 39
## Anderson Localization
Proceedings of the Fourth Taniguchi International Symposium, Sanda-shi, Japan, November 3–8, 1981
Editors: Y. Nagaoka, H. Fukuyama
1982. 98 figures. XII, 225 pages
ISBN 3-540-11518-8

Volume 38
## Physics of Intercalation Compounds
Proceedings of an International Conference Trieste, Italy, July 6–10, 1981
Editors: L. Pietronero, E. Tosatti
1981. 167 figures. IX, 323 pages
ISBN 3-540-11283-9

Volume 35
J. Bourgoin, M. Lannoo
## Point Defects in Semiconductors II
Experimental Aspects
With a Foreword by G. D. Watkins
1983. 116 figures. XVI, 295 pages
ISBN 3-540-11515-3

Volume 34
P. Brüesch
## Phonons: Theory and Experiments I
Lattice Dynamics and Models of Interatomic Forces
1982. 82 figures. XII, 261 pages
ISBN 3-540-11306-1

Volume 32
R. M. White
## Quantum Theory of Magnetism
2nd corrected and updated edition. 1983.
113 figures. XI, 282 pages
ISBN 3-540-11462-9
(Originally published by McGraw-Hill, Inc., New York, 1970)

Volume 30
M. Toda, R. Kubo, N. Saito
## Statistical Physics I
Equilibrium Statistical Mechanics
1983. 90 figures. XVI, 249 pages
ISBN 3-540-11460-2

Volume 29
## Electron Correlation and Magnetism in Narrow-Band Systems
Proceedings of the Third Taniguchi International Symposium, Mount Fuji, Japan, November 1–5, 1980
Editor: T. Moriya
1981. 99 figures. XIV, 257 pages
ISBN 3-540-10767-3

Volume 28
## The Structure and Properties of Matter
Editor: T. Matsubara
With contributions by T. Matsubara, H. Matsuda, T. Murao, T. Tsuneto, F. Yonezawa
1982. 223 figures. XI, 466 pages
ISBN 3-540-11098-4

Volume 27
S. V. Vonsovsky, Y. A. Izyumov, E. Z. Kurmaev
## Superconductivity of Transition Metals
Their Alloys and Compounds
Translated from the Russian by E. H. Brandt and A. P. Zavarnitsyn
1982. 182 figures. XIII, 512 pages
ISBN 3-540-11382-7

Springer-Verlag
Berlin
Heidelberg
New York
Tokyo

# Applied Physics A
Solids and Surfaces

**Applied Physics A** "Solids and Surfaces" is devoted to concise accounts of experimental and theoretical investigations that contribute new knowledge or understanding of phenomena, principles or methods of applied research.

Emphasis is placed on the following fields:

**Solid-State Physics**
Semiconductor Physics: **H. J. Queisser,** MPI Stuttgart
Amorphous Semiconductors: **M. H. Brodsky,** IBM Yorktown Heights
Magnetism (Materials, Phenomena): **H. P. Wijn,** Philips Eindhoven
Metals and Alloys, Solid-State Electron Microscopy: **S. Amelinckx,** Mol
Positron Annihilation: **P. Hautojärvi,** Espoo
Solid-State Ionics: **W. Weppner,** MPI Stuttgart

**Surface Science**
Surface Analysis: **H. Ibach,** KFA Jülich
Surface Physics: **D. Mills,** UC, Irvine
Chemisorption: **R. Gomer,** U. Chicago

**Surface Engineering**
Ion Implantation and Sputtering: **H. H. Andersen,** U. Aarhus
Laser Annealing: **G. Eckhardt,** Hughes Malibu
Integrated Optics, Fiber Optics,
Acoustic Surface Waves: **R. Ulrich,** TU Hamburg

Coordinating Editor: **H. K. V. Lotsch,** Heidelberg

**Special Features:**
- Rapid publication (3–4 months)
- No page charges for concise reports
- 50 complimentary offprints

**Subscription information** and/or **sample copies** are available from your bookseller or directly from Springer-Verlag, Journal Promotion Dept., P. O. Box 10 52 80, D-6900 Heidelberg, FRG

Springer-Verlag
Berlin
Heidelberg
New York
Tokyo